MW00770831

Quantum Dots

METHODS IN MOLECULAR BIOLOGY™

John M. Walker, SERIES EDITOR

METHODS IN MOLECULAR BIOLOGY™

Quantum Dots

Applications in Biology

Edited by

Marcel P. Bruchez
Charles Z. Hotz

Quantum Dot Corporation
Hayward, CA

HUMANA PRESS ✳ TOTOWA, NEW JERSEY

© 2007 Humana Press Inc.
999 Riverview Drive, Suite 208
Totowa, New Jersey 07512

www.humanapress.com

All rights reserved. No part of this book may be reproduced, stored in a retrieval system, or transmitted in any form or by any means, electronic, mechanical, photocopying, microfilming, recording, or otherwise without written permission from the Publisher. Methods in Molecular Biology™ is a trademark of The Humana Press Inc.

All papers, comments, opinions, conclusions, or recommendations are those of the author(s), and do not necessarily reflect the views of the publisher.

This publication is printed on acid-free paper. ∞
ANSI Z39.48-1984 (American Standards Institute) Permanence of Paper for Printed Library Materials.

Cover design by Patricia F. Cleary
Cover illustration: single quantum dot imaging provides a simple way to visualize the fraction of a cell membrane explored by tagged proteins (Fig. 3, Chapter 7; *see* complete caption on p. 88).
Production Editor: Jennifer Hackworth

For additional copies, pricing for bulk purchases, and/or information about other Humana titles, contact Humana at the above address or at any of the following numbers: Tel.: 973-256-1699; Fax: 973-256-8341; E-mail: orders@humanapr.com; or visit our Website: www.humanapress.com

Photocopy Authorization Policy:
Authorization to photocopy items for internal or personal use, or the internal or personal use of specific clients, is granted by Humana Press Inc., provided that the base fee of US $30.00 per copy is paid directly to the Copyright Clearance Center at 222 Rosewood Drive, Danvers, MA 01923. For those organizations that have been granted a photocopy license from the CCC, a separate system of payment has been arranged and is acceptable to Humana Press Inc. The fee code for users of the Transactional Reporting Service is: [978-1-58829-562-0 • 1-58829-562-1/07 $30.00].

Printed in Thailand by Imago. 10 9 8 7 6 5 4 3 2 1

eISBN: 1-59745-369-2
ISSN: 1064-3745

Library of Congress Cataloging-in-Publication Data

Quantum dots : biological applications / edited by Marcel Bruchez and Charles
Z. Hotz
    p. ; cm. -- (Methods in molecular biology, ISSN 1064-3745 ; 374
  Includes bibliographical references and index.
  ISBN-13: 978-1-58829-562-0
  ISBN-10: 1-58829-562-1 (alk. paper)
   1. Quantum dots. 2. Molecular biology. 3. Science. 1. Hotz, Charles Z.
II. Bruchez Marcel III. Series: Methods in molecular biology (Clifton,
N.J.) ; v. 374
   [DNLM: 1. Staining and Labeling--methods. 2. Fluorescent Dyes. 3.
Microscopy--methods. 4. Quantum Dots. W1 ME9616J v.374 / QH 237
Q23 2007]
   Q158.5.Q36 2007
   572'.36--dc22

                                              2006015499

Preface

Since the discovery of fluorescein more than a century ago, fluorescent dyes and stains have become workhorse reagents in experimental biology. Although much of this work has been accomplished with small-molecule dyes, there have been a number of notable exceptions (e.g., phycobiliproteins, expressible fluorescent proteins, and so on). However, although the number of small-molecule fluorescent probes has progressively increased, there have been few "inherently new" types of probes for use by the laboratory biologist.

Quantum dots are an inherently new class of fluorescent probes. Taking advantage of the unique attributes of nanometer-scale semiconductor particles, quantum dots (QDs) provide bright, stable, and sharp fluorescence for use in many biological labeling applications. Although small-molecule dyes, and even proteins, can be made as pure, homogeneous entities, quantum dots are engineered materials—consequently QDs from different sources may be different in composition and performance. This "materials" characteristic is important to consider, because the particular source for QDs may be vital to their functioning in a particular assay.

Since commercial debut of QDs only a few years ago, numerous publications have demonstrated their enabling utility in biological detection. *Quantum Dots: Applications in Biology* attempts to capture many of these diverse biological applications so the reader can not only observe what others have accomplished, but also anticipate new uses for these novel reagents.

This book is organized into five parts. The first two parts cover the use of QDs in imaging fixed and living cells (and tissues), respectively. Protocols are included for using QDs in routine (protein and structural cellular labeling), as well as in enabling applications (single-receptor trafficking, clinical pathology, correlative microscopy). Clearly, QDs will have a large impact in imaging applications owing to their outstanding photostability.

The third part shows early efforts aimed at using QDs in live animals. Although much work still needs to be done here, one again sees the promise of the technology to make a positive impact on human health.

The final two parts demonstrate the versatility of QD technology in existing assay technology. First, in flow cytometry, QDs show the promise of delivering more information—ultimately via a less expensive instrument platform. Additionally, one sees the application in immunosorbent (microplate) assays, reverse-phase protein arrays, and in a novel, multiplexed genotyping assay.

In early attempts to use QDs researchers were forced to develop their own protocols (which often differ dramatically from the analogous organic-dye protocols) to get even routine applications working. The original published procedures (usually available in the supplemental information along with primary publications) were not consistently or widely available, limiting substantially the number of people willing to try QDs in their applications. At this time, the published literature is growing very quickly. There have been over 250 peer-reviewed publications in this area since the original papers in 1998, and roughly half of these are now relying on commercially available reagents. Standard reagents and emerging standard protocols make quantum dots a well-established method for biological labeling in a wide variety of applications. The authors hope *Quantum Dots: Applications in Biology* provides a good starting reference for protocols that can be used to get reliable performance from QD-enabled applications.

Marcel P. Bruchez
Charles Z. Hotz

Contents

Contributors

BARNABY ABRAMS • *BD Biosciences, San Jose, CA*

RIZWAN S. AKHTAR • *Division of Neuropathology, Departments of Pathology and Neurobiology, University of Alabama at Birmingham, Birmingham, AL*

A. PAUL ALIVISATOS • *Materials Science Division, Department of Chemistry, University of California, Berkeley; Lawrence Berkeley National Laboratory, Berkeley, CA*

DONNA J. ARNDT-JOVIN • *Department of Molecular Biology, Max Planck Institute for Biophysical Chemistry, Goettingen, Germany*

MOUNGI G. BAWENDI • *Department of Chemistry, Massachusetts Institute of Technology, Cambridge, MA*

ROSANNE BOUDREAU • *Physical Bioscience Division, Lawrence Berkeley National Laboratory, Berkeley, CA*

CEDRIC BOUZIGUES • *Laboratoire Kastler Brossel, Physics Department, École Normale Supérieure, Paris, France*

MARCEL P. BRUCHEZ • *Quantum Dot Corporation, Hayward, CA*

PRATIP K. CHATTOPADHYAY • *Immunotechnology Section, Vaccine Research Center, National Institute of Allergy and Infectious Diseases, National Institutes of Health, Bethesda, MD*

LELAND W. K. CHUNG • *Department of Urology and Winship Cancer Institute, Emory University, Atlanta, GA*

MAXIME DAHAN • *Laboratoire Kastler Brossel, Physics Department, École Normale Supérieure, Paris, France*

THOMAS J. DEERINCK • *Department of Neurosciences, The National Center of Microscopy and Imaging Research, University of California, San Diego, La Jolla, CA*

TIM DUBROVSKY • *BD Biosciences, San Jose, CA*

MARK H. ELLISMAN • *Department of Neurosciences, The National Center of Microscopy and Imaging Research, University of California, San Diego, La Jolla, CA*

MAT FALKOWSKI • *Affymetrix Inc., South San Francisco, CA*

JOHN V. FRANGIONI • *Division of Hematology/Oncology, Department of Radiology, Beth Israel Deaconess Medical Center, Harvard Medical School, Boston, MA*

CARLO FUCCIO • *Department of Experimental Medicine, Section of Pharmacology, Second University of Naples, Naples, Italy*

XIAOHU GAO • *Department of Bioengineering, University of Washington, Seattle, WA*

DAVID H. GEHO • *Center for Applied Proteomics and Molecular Medicine, George Mason University, Manassas, VA*

BEN N. G. GIEPMANS • *Department of Neurosciences, The National Center of Microscopy and Imaging Research, University of California, San Diego, La Jolla, CA*

ELLEN R. GOLDMAN • *Center for Bio/Molecular Science and Engineering, Naval Research Laboratory, Washington, DC*

WEIWEI GU • *Department of Anatomy, University of California, San Francisco, San Francisco, CA*

PREM GURNANI • *Division of Translational Pathology, Department of Pathology, University of Texas Southwestern Medical Center, Dallas, TX*

ANDREW HAYHURST • *Department of Virology and Immunology, Southwest Foundation for Biomedical Research, San Antonio, TX*

CHARLES Z. HOTZ • *Quantum Dot Corporation, Hayward, CA*

JOHBU ITOH • *Teaching and Research Support Center, Division of Cell Science, Tokai University School of Medicine, Kanegawa, Japan*

MANEESH JAIN • *Affymetrix Inc., South San Francisco, CA*

JYOTI K. JAISWAL • *Department of Cellular Biophysics, The Rockefeller University, New York, NY*

THOMAS M. JOVIN • *Department of Molecular Biology, Max Planck Institute for Biophysical Chemistry, Goettingen, Germany*

GEORGE KARLIN-NEUMANN • *Affymetrix Inc., South San Francisco, CA*

J. KEITH KILLIAN • *FDA–NCI Clinical Proteomics Program, Laboratory of Pathology, National Cancer Institute, National Institutes of Health, Bethesda, MD*

SANG-WOOK KIM • *Department of Chemistry, Massachusetts Institute of Technology, Cambridge, MA*

SUNGJEE KIM • *Department of Chemistry, Massachusetts Institute of Technology, Cambridge, MA*

JOAN H. M. KNOLL • *Department of Pediatrics, Children's Mercy Hospital, University of Missouri-Kansas City School of Medicine, Kansas City, MO*

B. CHRISTOFFER LAGERHOLM • *Department of Cell and Developmental Biology, University of North Carolina, Chapel Hill, NC; presently, MEMPHYS, University of Southern Denmark, Odense, Denmark*

CAROLYN A. LARABELL • *Department of Anatomy, University of California, San Francisco; Physical Bioscience Division, Lawrence Berkeley National Laboratory, University of California, San Francisco, San Francisco, CA*

CECELIA B. LATHAM • *Division of Neuropathology, Departments of Pathology and Neurobiology, University of Alabama at Birmingham, Birmingham, AL*

MARK A. LE GROS • *Physical Bioscience Division, Lawrence Berkeley National Laboratory, Berkeley, CA*

SABINE LÉVI • *Laboratoire de Biologie Cellulaire de la Synapse, Biology Department, École Normale Supérieure, Paris, France*

DIANE S. LIDKE • *Department of Pathology, University of New Mexico, Albuquerque, NM*

STEVEN LIN • *Affymetrix Inc., South San Francisco, CA*

HONGJIAN LIU • *ReLIA Diagnostic Systems, LLC, Burlingame, CA*

MARYANN E. MARTONE • *The National Center of Microscopy and Imaging Research, Department of Neurosciences, University of California, San Diego, La Jolla, CA*

HEDI MATTOUSSI • *Division of Optical Sciences, Naval Research Laboratory, Washington, DC*

OSMAN MUHAMMAD • *Brain Tumor Institute, Cleveland Clinic Foundation, Cleveland, OH*

PETER NAGY • *Department of Biophysics and Cell Biology, Medical and Health Science Center, University of Debrecen, Debrecen, Hungary*

ANIMESH NANDI • *Division of Translational Pathology, Department of Pathology, University of Texas Southwestern Medical Center, Dallas, TX*

SHUMING NIE • *Departments of Biomedical Engineering, Chemistry, Hematology, and Oncology, and Winship Cancer Institute, Emory University and Georgia Institute of Technology, Atlanta, GA*

SHUNSUKE OHNISHI • *Division of Hematology/Oncology, Department of Radiology, Beth Israel Deaconess Medical Center, Harvard Medical School, Boston, MA*

RICHARD L. ORNBERG • *Alcon Laboratories, Fort Worth, TX*

ROBERT YOSHIYUKI OSAMURA • *Department of Pathology, Tokai University School of Medicine, Kanegawa, Japan*

WOLFGANG J. PARAK • *Department of Chemistry, University of California, Berkeley, Berkeley, CA*

JOHANNE PASTOR • *Division of Translational Pathology, Department of Pathology, University of Texas Southwestern Medical Center, Dallas, TX*

TERESA PELLEGRINO • *Department of Chemistry, University of California, Berkeley, Berkeley, CA*

ALEXANDRA POPESCU • *Brain Tumor Institute, Cleveland Clinic Foundation, Cleveland, OH*

MARIO ROEDERER • *Immunotechnology Section, Vaccine Research Center, National Institute of Allergy and Infectious Disease, National Institutes of Health, Bethesda, MD*

KEVIN P. ROSENBLATT • *Division of Translational Pathology, Department of Pathology, University of Texas Southwestern Medical Center, Dallas, TX*

KEVIN A. ROTH • *Division of Neuropathology, Departments of Pathology and Neurobiology, University of Alabama at Birmingham, Birmingham, AL*

MARINA SEDOVA • *Affymetrix Inc., South San Francisco, CA*

SANFORD M. SIMON • *Department of Cellular Biophysics, The Rockefeller University, New York, NY*

DARIO SINISCALCO • *Department of Experimental Medicine, Section of Pharmacology, Second University of Naples, Naples, Italy*

BENJAMIN L. SMARR • *The National Center of Microscopy and Imaging Research, Department of Neurosciences, University of California, San Diego, La Jolla, CA*

STEVEN A. TOMS • *Brain Tumor Institute, Cleveland Clinic Foundation, Cleveland, OH*

ANTOINE TRILLER • *Laboratoire de Biologie Cellulaire de la Synapse, Biology Department, École Normale Supérieure, Paris, France*

H. TETSUO UYEDA • *Division of Optical Sciences, Naval Research Laboratory, Washington, DC*

ZHIYONG WANG • *Affymetrix Inc., South San Francisco, CA*

PAUL G. WYLIE • *TTP LabTech Ltd., Melbourn, Royston, Hertfordshire, UK*

JOANNE YU • *Immunotechnology Section, Vaccine Research Center, National Institute of Allergy and Infectious Disease, National Institutes of Health, Bethesda, MD*

I

IMAGING OF FIXED CELLS AND TISSUES

1

Immunofluorescent Labeling of Proteins in Cultured Cells With Quantum Dot Secondary Antibody Conjugates

Richard L. Ornberg and Hongjian Liu

Summary

Our understanding of the level and distribution of gene and protein expression in cells is a key component of modern cell biological and medical research. Detecting intracellular proteins with labeled antibodies or genes with labeled oligonucleotide sequences by fluorescence microscopy requires fixation of the target molecule in its natural distribution and the penetration of the probe to the target. This typically involves chemical fixation followed by a detergent treatment that renders the cell membrane permeable to a labeling antibody. The advantages of using quantum dots (QDs) over organic dyes to detect expression, such as high brightness, stability, and simplified multiple target labeling has been described in previous publications. However, QDs are structurally larger than organic dye probes and require different fixation and permeabilization conditions for optimum labeling. In the chapter, we describe several protocols for labeling proteins in nuclear, cytoplasmic, and membranous compartments with QD conjugates.

Key Words: Quantum dot; light microscopy, fluorescence microscopy, immunolabeling, multiplex labeling.

1. Introduction

Driven by the desire to systematically characterize cell function and pathways and the availability of new antibodies, labeling protocols for cells are continually being developed to measure cellular distribution, level of expression and, most recently, biochemical status, i.e., phosphorylation state, of an ever-growing number proteins (*1,2*). Labeling for protein expression typically begins with a chemical treatment that either crosslinks proteins with aldehydes, or precipitates them with organic solvents such as cold acetone or methanol. The choice of fixative is largely dependent on the chemical sensitivity of the epitope on the

From: *Methods in Molecular Biology, vol. 374: Quantum Dots: Applications in Biology*
Edited by: M. P. Bruchez and C. Z. Hotz © Humana Press Inc., Totowa, NJ

protein to be labeled and/or the need to fix the cells quickly in order to capture transient distributions. Fixation is followed by permeabilization treatments with detergents to remove the membranous and other molecular barriers to labeling antibodies. The choice of detergent is determined by the degree of permeabilization required. Typically, mild natural detergents such as saponin and nonionic detergents such as Triton X-100 or NP-40 are used at low concentration in buffered saline to solubilized membrane lipids and nonfixed protein material. Recently the cationic detergent, dodecyl trimethyl ammonium chloride (DTAC), has been promoted to preserve cytoskeletal proteins distributions in fixed cells *(3,4)*. The choice of detergent is empirically determined by the ability to label the sites of interest and the preservation of the cellular structures of interest.

Regardless of the choice of reagents, the fixation and permeabilization of a cell compromises the natural distribution and physiological state of proteins in a cell. The goal is to capture targets and provide access to labeling reagents while disturbing the natural distribution as little as possible. Consequently, each labeling protocol is a judicious compromise and optimization of reagents, concentration, temperature, and time for any given cell and protein.

The protocols that follow describe the labeling of proteins with antibodies and secondary antibody quantum dot (QD) conjugates in acetone-methanol fixed (**Fig. 1**), or paraformaldehyde-fixed cultured cells (**Figs. 2** and **3**). They are provided as starting points for QD-labeling experiments.

2. Materials

2.1. Fixatives

1. 16% Paraformaldehyde solution, EM grade (Electron Microscopy Sciences, Inc., Hatfield, PA): prepare 50 mL 2% paraformaldehyde fixative by diluting 6.25 mL 16% solution and 5.0 mL 10 mM PBS with 38.75 mL distilled water.
2. Methanol–acetone fixative: make 100 mL of fixative by adding 10 mL 20 mM EGTA solution, 70 mL methanol, and 20 mL acetone into a clean glass bottle, and storing it at –20°C for at least 2 h prior to use. This solution can be stored at –20°C for reuse at a later date.

2.2. Buffers

1. 10X Phosphate-buffered saline (PBS) solution, pH 7.2 (Sigma-Aldrich, St. Louis, MO): prepare PBS by diluting 1 part 10X PBS stock into 9 parts distilled H$_2$O.
2. Tris-buffered saline (TBS): 10 mM Tris-HCl, 130 mM NaCl, pH 7.4.
3. Borate buffer: 50 mM sodium borate, pH 8.3.
4. EGTA (Sigma, cat. no. E-3889): prepare 0.1 M solution of EGTA by dissolving 3.8 g of EGTA in water. Adjust pH to 7.0–8.0 with 1 N NaOH.
5. Permeabilization buffers:
   a. 2% (v/v) Triton X-100 in TBS (Sigma-Aldrich T-9284).
   b. 0.5–2.0% (w/v) DTAC in TBS (Sigma-Aldrich 44242).

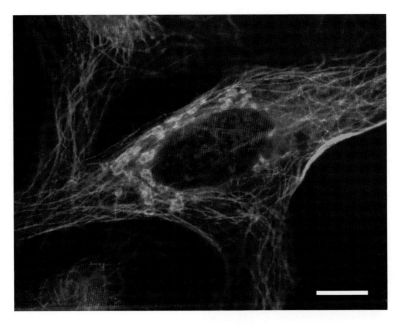

Fig. 1. Light micrograph of a HeLa cell fixed with cold methanol/acetone and labeled with quantum dot (QD) secondary antibody conjugates. Microtubules (green) were labeled with a rat anti-tubulin antibody and a 525-nm QD anti-rat secondary antibody conjugate. The Golgi apparatus was labeled with a rabbit anti-giantin antibody followed by a 585-nm anti-rabbit secondary antibody conjugate. The nucleus was labeled with a mouse anti-nucleosome antibody followed by a 655-nm QD anti-mouse secondary antibody conjugate. Bar = 5 µ.

6. Rinse buffer: 0.05% (v/v) Tween-20 in TBS (Dako Cytomation S3304 or equivalent).
7. Blocking buffer: bovine serum albumin (BSA), (IgG-free BSA; Jackson Laboratories 001-000-162) or equivalent. Prepare 2% BSA and 0.05% Tween-20 in TBS.

2.3. Antibodies

1. Mouse anti-nucleosome antibody (BD Pharmingen, San Jose, CA, cat. no. 550869).
2. Rat anti-α-tubulin antibody (Serotech, Raleigh, NC, cat. no. MCA77G).
3. DSB-X biotinylated goat anti-rat IgG (Molecular Probes, Eugene, OR, cat. no. D-20697).
4. Rabbit anti-giantin antibody (Covance Research Products, Denver, PA, cat. no. PRB-114C).
5. Rabbit anti-phosphorylated mitogen activated protein kinase (pMAPK) (Cell Signaling Technology, Danvers, MA, cat. no. 4377).

2.4. QD Conjugates

1. Qdot® 655 anti-rabbit secondary antibody conjugate (Quantum Dot Corp., Hayward, CA, cat. no. 1142-1).

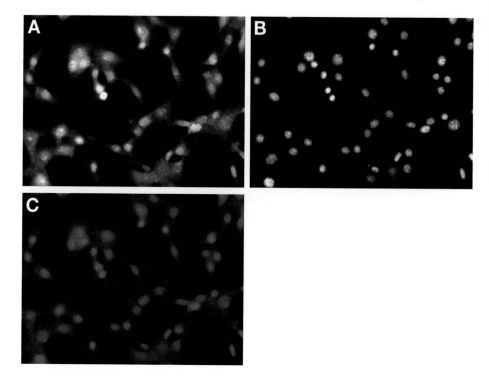

Fig. 2. Light micrograph of nuclear phosphorylated p42/44 mitogen-activated protein kinase (pMAPK) **(A)** and nuclei **(B)** in paraformaldehyde-fixed NIH 3T3 cells. Cells were serum starved for 6 h and stimulated with 50 μg/mL platelet-derived growth factor for 10 min prior to fixation. pMAPK was labeled with a rabbit anti-pMAPK primary antibody followed by a 655-nm quantum dot anti-rabbit secondary antibody conjugate. Nuclei were labeled with Hoechst 33342. **(C)** Merged images of **A** and **B** illustrating the nuclear translocation of pMAPK following growth factor stimulation.

2. Qdot 525 streptavidin conjugate (Quantum Dot Corp., cat. no. 1014-1).
3. Qdot 585 anti-rabbit secondary antibody conjugate (Quantum Dot Corp., cat. no. 1141-1).
4. Qdot 655 anti-mouse secondary antibody conjugate (Quantum Dot Corp., cat. no. 1102-1).

2.5. Qdot Conjugate Conjugate Buffers

1. 2% BSA in 0.1 *M* borate buffer containing 5–20 n*M* (1:200–1:50) Qdot® secondary antibody conjugate (Quantum Dot Corp.).

2.6. Nuclear Stains

1. For red-emitting QDs (Qdot 605, 655, 705, 800); Hoechst 33342 nuclear stain (Molecular Probes) diluted to 0.5 μg/mL (1:20,000) in TBS-Tween.

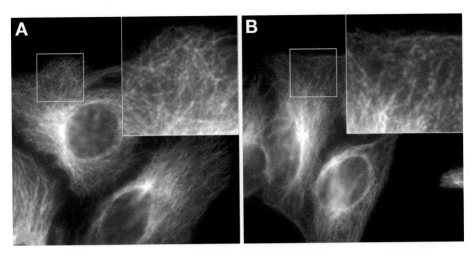

Fig. 3. Images of microtubule labeling in paraformaldehyde-fixed HeLa cells fixed using 4% paraformaldehyde and permeabilized in 0.2% Triton X-100. **(A)** Tubulin labeled with rat anti-tubulin antibody and 565 Qdot anti-rat secondary antibody conjugate. **(B)** Tubulin labeled with rat anti-tubulin antibody followed by Alexa 546 anti-rat secondary antibody conjugate. The insets provide a high magnification comparison of the quantum dot and organic dye-labeling quality for the cytoplasmic protein.

2. For green-to-orange-emitting QDs (Qdot 525, 565, 585, 605); DRAQ5 DNA probe (AXXORA LLC, San Diego, CA) diluted 1:500 in TBS-Tween.

2.7. Mounting Medium Reagents

1. Polyvinyl alcohol (PVA)–DABCO mounting medium (Sigma–10981; PVA mounting medium with DABCO antifading agent).
2. Glycerol in borate: 90% glycerol in 50 mM borate buffer, pH 8.0–8.3.
3. Ethanol: 100% ACS reagent grade.
4. Toluene: American Chemical Society (ACS) reagent grade.
5. Tri-*n*-octyl phosphine (TOP) (Fisher Scientific ICN222338): prepare 10% solution in toluene.
6. Cytoseal 60™-mounting medium (Richard-Allan Scientific, Kalamazoo, MI, cat. no. 8310) containing 1.0% TOP. Dilute 10% TOP-toluene solution 1:10 in Cytoseal 60. Mix gently to avoid introduction of air bubbles (*see* **Note 1**).

3. Methods

3.1. Fixation

3.1.1. Methanol-Acetone Fixation

1. Rinse the cells with ice-cold PBS, two times for 5 min (*see* **Note 2**). Remove excess buffer and put culture dish on ice.

2. Immediately fix cells with cold (–20°C) fixative and put culture dish into a –20°C freezer for 25 min.
3. Remove cells from freezer and wash with room temperature TBS three times for 3 min each. Proceed to labeling.

3.1.2. Formaldehyde Fixation (see **Note 2**)

1. Rinse cells briefly with ice-cold PBS or compatible saline to remove culture medium protein.
2. Fix cells at room temperature in paraformaldehyde/PBS for 10 min.

3.2. Permeabilization

3.2.1. Nuclear Protein Labeling

1. Fix cells in 1% paraformaldehyde in PBS or TBS for 10 min.
2. Wash cells in TBS; three changes for 5 min each.
3. Incubate cells in 2.0% DTAC permeabilization buffer for 20 min.
4. Wash in three changes of TBS saline for 5 min each.

3.2.2. Cytoplasmic and Membrane Protein Labeling

1. Fix cells in 1–2% paraformaldehyde in PBS or TBS for 10 min.
2. Wash cells in TBS; three changes for 5 min each.
3. Incubate cells in 0.1–0.2% Triton X-100 in TBS for 10 min.
4. Wash cells in TBS; three changes for 5 min each.

3.3. Labeling Cellular Proteins in Cultured Cells

1. Briefly rinse cells in blocking buffer.
2. Incubate in blocking buffer for 20 min.
3. Remove excess blocking buffer.
4. Incubate in primary antibodies in blocking buffer from 1 h to overnight at 2–8°C (refrigerator temperature).
5. Rinse cells in three changes of TBS-Tween rinse buffer for 3 min each.
6. Rinse cells once in borate buffer briefly and block for 15 min in 2% BSA-borate buffer.
7. Incubate cells in Qdot Secondary conjugate solution for 60 min at room temperature.
8. Wash cells in three changes of borate buffer for 5 min each.
9. Incubate in nuclear staining solutions for 5 min.
10. Wash in borate buffer.

3.4. Specimen Mounting (see **Note 3**)

3.4.1. Aqueous-Based Mounting Procedure

Mounting labeled subsequent imaging can be done with a variety of water-based mounting mediums.

1. Labeled cells are mounted by placing a small drop (10 μL for a 22 × 22-mm cover slip) of glycerol-borate or PVA-DABCO mounting medium on the cells and carefully placing a cover slip on the solution.
2. Preparation mounted in glycerol-borate with cover slips should be sealed around the edges with clear finger polish.

3.4.2. Nonaqueous Mounting Procedure

Nonaqueous mounting is an alternative method for cells on substrate that can withstand organic solvents, i.e., glass. This mounting medium has the advantage of preserving QD luminescence for very long periods of time.

1. Dehydrate cells in a grade series of ethanol/water solutions such as 50, 70, 90, 100, 100, 100% (v/v).
2. Rinse in 100% toluene.
3. Apply a drop of Cytoseal 60-TOP mixture to cells and cover slip.

4. Notes

1. Qdot conjugates can be mounted in either aqueous or nonaqueous mounting medium. Aqueous mounting can be done in solutions of glycerol and/or PVA (Gelvatol, Mowiol). In the case of glycerol-mounting medium, high concentrations, i.e., 90%, in buffer have been reported to be better than 50% glycerol solutions (*5*). Aqueous mounting should be used with cells that have been labeled with both organic dyes and Qdot conjugates to preserve the fluorescence of the organic dye. These have the advantage of preserving cellular structures for phase-contrast imaging and are typically required to extend organic dye fluorescence. Such media do not necessarily preserve QD fluorescence however. Several media have been used for QD-labeled cells including glycerol in the buffer, and mixtures of PVA/glycerol and buffer such as the PVA-DABCO reagent listed in **Subheading 2.7., item 1**. The latter are more permanent in that they harden to firmly affix a cover slip to a slide. Specimens mounted in solutions of glycerol are more fragile and require sealing the outer edges of the cover slip with nail polish.
2. As with all immunolabeling experiments, optimization of fixation and permeabilization conditions must be performed for each primary antibody and cell type. Labeling nuclear proteins with QD conjugates is very sensitive to overfixation. Aldehyde fixation is known to induce cell shrinkage and changes in nuclear structure (*6*). Formaldehyde/paraformaldehyde concentrations greater than 2% or fixation times longer than 10 min drastically reduce nuclear labeling with Qdot secondary sntibody conjugates, (unpublished observations). Nuclear labeling with Qdot secondary antibody conjugates requires the use of 2% DTAC as the permeabilization reagent.
3. Several organic-mounting mediums are available, however, many have a green autofluorescence after drying. Cytoseal 60 appears to have essentially no autofluorescence when irradiated with excitation light appropriate for QDs. The addition of TOP, a ligand used in the synthesis of QDs, to the Cytoseal medium enhances and stabilizes dot luminescence. TOP is liquid at room temperature and is particularly

sensitive to oxygen. The mounting medium should be prepared fresh and the stock TOP solution should be kept sealed when not in use.

References

1. Larsson, L. I. (1988) *Immunocytochemistry: Theory and Practice.* CRC Press, Boca Raton, FL.
2. Jacobberger, J. W. (1991) Intracellular antigen staining: quantitative immunofluorescence. *Methods* **2,** 207–218.
3. Nakamura, F. (2001) Biochemical, electron microscopic and immunohistological observations of cationic detergent-extracted cells: detection and improved preservation of microextensions and ultramicroextensions. *BMC Cell Biol.* **2,** 10.
4. Callow, M. G., Zozulya, S., Gishizky, M. L., Jallal, B., and Smeal, T. (2005) PAK4 mediates morphological changes through the regulation of GEF-H1. *J. Cell Science* **118,** 1861–1872.
5. Ness, J. M., Akhtar, R. S., Latham, C. B., and Roth, K. A. (2003) Combined tyramide signal amplification and quantum dots for sensitive and photostable immunofluorescence detection. *J. Histochem. Cytochem.* **51,** 981–987.
6. Kozubeka, S., Lukáováa, E., Amrichováa, J., Kozubekb, M., Likováa, A., and lotováa, J. (2000) Influence of cell fixation on chromatin topography. *Anal. Biochem.* **282,** 29–38.

2

Immunohistochemical Detection With Quantum Dots

Rizwan S. Akhtar, Cecelia B. Latham, Dario Siniscalco, Carlo Fuccio, and Kevin A. Roth

Summary

Quantum dot (QD) conjugates have many immunohistochemical applications. The optical, excitation/emission, and photostable properties of QDs offer several advantages over the use of chromogens or organic fluorophores in these applications. Here, we describe the use of QD conjugates to detect primary antibody binding in fixed tissue sections. We also describe the use of QDs in simultaneous and sequential multilabeling procedures and in combination with enzyme-based signal amplification techniques. QD conjugates expand the arsenal of the immunohistochemist and increase experimental flexibility in many applications.

Key Words: Immunostaining; detection methods; multi-labeling; tyramide signal amplification; fluorophore; antibody binding.

1. Introduction

Detection of biologically relevant molecules in human and animal fixed-tissue sections is often accomplished by applying primary antibodies that bind specifically to the antigen of interest followed by application of labeled secondary antibodies. Traditionally, the secondary antibodies used in immunohistochemical detection have been directly conjugated to organic fluorophores. These labels are easily visualized using a suitably equipped fluorescence microscope. Alternatively, the secondary antibodies may not be labeled *per se*, but rather conjugated to enzymes that catalyze the deposition of chromogenic or fluorescent substrates. For example, horseradish peroxidase (HRP) or alkaline phosphatase (AP)-conjugated secondary antibodies can be used to deposit fluorescent or chromogenic substrates at and/or near the location of primary antibody binding. In this manner, enzyme-based amplification can greatly increase the sensitivity of immunohistochemical detection *(1)*.

From: *Methods in Molecular Biology, vol. 374: Quantum Dots: Applications in Biology*
Edited by: M. P. Bruchez and C. Z. Hotz © Humana Press Inc., Totowa, NJ

Quantum dot (QD) conjugates have recently become commercially available (as Qdot® Conjugates from Quantum Dot Corporation, Hayward, CA) and can be used in several immunohistochemical applications. These reagents offer several advantages over typical fluorophores. First, many organic fluorophores undergo a rapid and irreversible photobleaching. QDs photobleach minimally, if at all, even with extended periods of viewing *(2–5)*. Second, all QDs have a similar excitation spectrum and can be excited in the near 360-nm range. This feature, coupled with each QD narrow emission spectra, facilitates their use in simultaneous multilabeling techniques *(2,6)*. QD-based immunohistochemistry can allow multilabeling visualization without the need for computer image overlay or multiple custom filtersets, one or both of which is typically required with organic fluorophores.

In this chapter, we describe how to use QD conjugates in several immunohisto-chemical applications. We demonstrate multilabeling QD protocols and how to combine QD conjugates with tyramide signal amplification (TSA) for exquisitely sensitive immunohistochemical detection. Finally, we describe how to combine these techniques to accomplish simultaneous triple labeling. For completeness, we begin with an overview of our tissue section preparation protocol.

2. Materials

2.1. Preparation of Tissue for Immunohistochemical Study

1. 10X Phosphate buffered saline (PBS): dissolve 80 g NaCl, 2.0 g KCl, 14.2 g Na_2HPO_4, 2.0 g KH_2PO_4, and 0.1 g NaN_3 in 1 L distilled H_2O. Adjust pH to 7.2 with NaOH. Store 10X solution at room temperature. Dilute 10X using distilled H_2O for 1X PBS and store at room temperature.
2. 4% (w/v) Paraformaldehyde: dissolve 4 g paraformaldehyde (Electron Microscopy Sciences, Fort Washington, PA; cat. no. 19200) in PBS, heat to 58–60°C, and add 1 N NaOH dropwise until solution is clear. Fill to 100 mL volume with PBS and pH to 7.2. Use immediately. If solution is heated above 60°C, do not use.
3. Bouin's fixative: 750 mL saturated aqueous picric acid (Sigma, St. Louis, MO, cat. no. P-6744), 250 mL formaldehyde solution (Fisher Scientific, Pittsburgh, PA, cat. no. F75-1GAL), and 50 mL glacial acetic acid (Fisher Scientific, cat. no. A58-500). Store at 4°C.
4. Tissue-freezing medium (VWR International, West Chester, PA, cat. no. 15148-031).
5. Slides: Snowcoat X-tra (Surgipath Medical Industries, Inc., Richmond, IL, cat. no. 15148-031).
6. Slide rack/staining dish: Wheaton 900200 (VWR International, cat. no. 25461-003).
7. Antigen retrieval container: TPX staining jar (VWR International, cat. no. 25460-907).

8. CitriSolv (Fisher Scientific, cat. no. 22-143975).
9. Isopropanol (Fisher Scientific, cat. no. HC-500-1GAL).
10. Citrate antigen retrieval buffer: working solution (10 mM citrate buffer, pH 6.0) consists of 9 mL 0.1 M citric acid monohydrate solution (Sigma, cat. no. C-7129), 41 mL 0.1 M sodium citrate solution (Sigma, cat. no. S-4641), and 450 mL distilled H_2O for a total of 500 mL. Store all solutions at 4°C.
11. Steamer: Farberware FRA500 (domestic rice cooker) or equivalent.

2.2. Detection of Single Antibody Immunoreactivity With QD-Conjugated Secondary Antibody

1. Plastic slide folder (VWR International, cat. no. 48443-850).
2. Pap pen (Research Products International, Mount Prospect, IL; cat. no. 195504) was used in this study. Quantum Dot Corporation recommends the use of the ImmEdge Hydrophobic Barrier Pen (Vector Labs, Burlingame, CA; cat. no. H-4000).
3. PBS-blocking buffer (BB): dissolve 1.0 g bovine serum albumin (Fisher Scientific, cat. no. BP1600-100), 0.2 g nonfat powdered skim milk (Carnation, Nestle, Glendale, CA), and 0.3 mL Triton X-100 (Sigma, cat. no. T-8787) in 100 mL 1X PBS. Store at 4°C.
4. Bisbenzimide (Hoescht 33,258) stock solution: prepare Hoescht 33,258 (Sigma, cat. no. B-2883) at 2 mg/mL stored at 4°C. Dilute 1 µL in 10 mL PBS when applying to slides.
5. 90% (v/v) Glycerol in PBS (Fisher Scientific, cat. no. G33-1).
6. Microscope cover glasses: various sizes (VWR International, cat. no. 48393).
7. Optical fluorescence filters: usable excitation filters include 325 ± 25 nm (UV), 360 ± 40 nm (DAPI), 470 ± 40 nm (FITC), or 545 ± 30 nm (rhodamine). Usable emission filters include 400 nm long pass, 450 ± 50 nm (DAPI), 535 ± 40 nm (FITC), or 610 ± 75 nm (Cy3). Custom filtersets for QDs are available from Chroma Technology (Rockingham, VT) or Omega Optical (Brattleboro, VT).

2.3. Simultaneous Detection of Two Antibodies Raised in Different Species Using QD-Conjugated Secondary Antibodies

1. Mouse anti-MAP2 primary monoclonal antibody (Sigma, cat. no. M4403).
2. Rabbit anti-GFAP primary antiserum (Dako, Glostrup, Denmark, cat. no. Z334).
3. Quantum dot 565 anti-rabbit secondary antibody: QD 565 goat F(ab')2 anti-rabbit IgG conjugate (Quantum Dot Corp., cat. no. 1143-1).
4. QD 605 anti-mouse secondary antibody: QD 605 goat F(ab')2 anti-mouse IgG conjugate (Quantum Dot Corp., cat. no. 1100-1).

2.4. Detection of Single Antibody Immunoreactivity by TSA and a QD-Conjugated Streptavidin

1. 3% Hydrogen peroxide: dilute stock 30% hydrogen peroxide (Fisher Scientific, cat. no. BP2633) in PBS at time of use. Store stock solution at 4°C.

2. HRP-conjugated secondary antibody: peroxidase AffiniPure™ donkey anti-rabbit IgG (cat. no. 711-035-152) and peroxidase AffiniPure donkey anti-mouse IgG (cat. no. 715-035-151) both from Jackson ImmunoResearch Laboratories, West Grove, PA.
3. TSA™ Biotin System (Perkin-Elmer Life Sciences Products, Boston, MA; cat. no. NEL700).
4. QD 525 streptavidin conjugate (Quantum Dot Corp., cat. no. 1014-1).
5. QD 605 streptavidin conjugate (Quantum Dot Corp., cat. no. 1010-1).

3. Methods

3.1. Preparation of Tissue for Immunohistochemical Study

1. Sample tissue should be harvested and immediately fixed in a sufficient volume of fixative, such as 4% paraformaldehyde, Bouin's fixative, or methanol (*see* **Note 1**). In general, overnight immersion fixation at 4°C is satisfactory; however, for some antigens, perfusion fixation and/or special fixation procedures may be required for optimal immunohistochemical detection.
2. Sample tissue can be prepared as either paraffin embedded or frozen sections. Paraffin-embedded tissue sections require deparaffinization prior to use. Frozen sections should be allowed to warm to room temperature (approx 10 min at room temperature) prior to further processing. Go to **step 6** for frozen slides.
3. For paraffin sections, deparaffinize sections by washing slides twice in CitriSolv, first for 10 min and second for 5 min (*see* **Notes 2–4**). Then, rehydrate sections by three washes with isopropanol for 5 min each. Then, wash slides in running tap water for 5 min. Finally, wash slides in distilled water for 5 min.
4. To increase the immunohistochemical detection of some antigenic epitopes in fixed tissue, antigen retrieval (AR) may be required (*see* **Note 1**). We typically use a citric acid AR method. Perform AR by placing slides in an AR container containing citrate buffer. Place the container inside a steamer with an adequate amount of distilled water and turn it on. Once steam can be seen exiting the steamer, continue AR for 20 min. Then, remove the slide container, place it on the lab bench, and allow it to cool at room temperature for 20 min.
5. Remove slides from AR container and wash in running tap water for 5 min.
6. Incubate sections in PBS for 5 min at room temperature. Slides are now ready for immunostaining.

3.2. Detection of Single Antibody Immunoreactivity With QD-Conjugated Secondary Antibody

Direct immunofluorescence can be used to detect primary antibody binding in fixed-tissue sections. Many protocols use organic fluorophore-conjugated secondary antibodies that recognize species-specific epitopes on primary antibodies. QD-conjugated secondary antibodies can replace organic fluorophore-conjugated secondary antibodies in a number of immunohistochemical applications. An

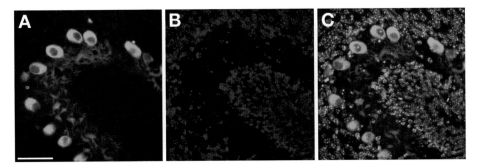

Fig. 1. Detection of single antibody immunoreactivity with quantum dot (QD)-conjugated secondary antibody. **(A–C)** Immunostained section of postnatal day 7 mouse cerebellum. **(A)** Calbindin, a marker of Purkinje cells in the cerebellum, is immunostained with a rabbit polyclonal antiserum, subsequently detected with QD 565-conjugated goat anti-rabbit secondary antibody and imaged with a 470 ± 40-nm excitation filter and a 535 ± 40-nm emission filter. **(B)** Hoescht 33,258 counterstaining of nuclei is imaged with a 360 ± 40-nm excitation filter and a 450 ± 50-nm emission filter. **(C)** Calbindin and Hoescht 33,258 counterstaining is simultaneously visualized with a 325 ± 25-nm excitation filter and a 400LP emission filter. Scale bar = 40 μm.

example of single antibody immunoreactivity detection with QD-conjugated secondary antibody is shown in **Fig. 1**.

1. Follow **Subheading 3.1.** to prepare slides (*see* **Note 5**).
2. Prepare a humidified chamber by cutting a paper towel in half lengthwise and placing it in a plastic slide folder. Wet the paper towel with distilled water. The paper towel should be damp but not completely saturated with water and should be flat. When slides are placed in the folder, the underside of the slides should not come into contact with the paper towel.
3. Quickly blot excess moisture from around the tissue on each slide with a paper towel and encircle the tissue with a Pap pen or equivalent (*see* **Note 6**). Place the slides in the humidified chamber and immediately add PBS-BB on top of the tissue within the Pap pen boundary. One slide should be processed at a time and care should be taken to avoid drying of the tissue.
4. To inhibit nonspecific antibody binding, incubate the slides in PBS-BB (blocking buffer) for 30 min at room temperature. As mentioned in **step 3**, excess PBS-BB can be added to individual slides one at a time. Alternatively, slides can be processed in bulk by placing the slides in a slide rack and immersing the rack in a staining dish containing PBS-BB.
5. Prepare primary antibody diluted in PBS-BB. Remove the PBS-BB from the slide (*see* **Note 7**) and apply primary antibody. When applying any antibody or reagent to slides, use a sufficient volume to cover the tissue (typically 50–200 μL).

6. Incubate the slides in primary antibody for 24 h at 4°C (*see* **Note 8**). Incubation can be shortened to 12 h at 4°C or to 60–90 min at room temperature. In general, longer antibody incubations at lower temperatures result in better signal-to-noise ratios in immunohistochemical detection.
7. Wash the slides three times in PBS for 5 min each (*see* **Note 9**).
8. Dilute QD-conjugated secondary antibody in PBS-BB at a 1:100 dilution. Because the Pap pen may interfere with QD signals *(7)*, remove the Pap pen from slides before applying QD conjugates (*see* **Note 6**). Apply secondary antibody and incubate slides in a humidified chamber for 1 h at room temperature.
9. Wash the slides three times in PBS for 5 min each. If nuclear counterstaining is desired, dilute stock bisbenzimide (Hoescht 33,258) solution in the PBS during the second wash.
10. Cover slip slides in 90% glycerol in PBS (*see* **Note 10**).
11. View slides with a fluorescence microscope equipped with a 360 ± 40 nm (DAPI) excitation filter and a 400-nm long pass emission filter (*see* **Note 11**) (*see* **Subheading 3.7.** for information on optimal filterset selection).

3.3. Simultaneous Detection of Two Antibodies Raised in Different Species Using Two QD-Conjugated Secondary Antibodies

There are several variations and extensions of the protocol in **Subheading 3.2.** that will allow the detection of multiple antigens in fixed tissue. Multiple primary antibodies can be applied simultaneously to a single tissue section. After an overnight incubation period, multiple secondary antibodies can then be applied to detect primary antibody binding. As long as each primary antibody is raised in a different species, this technique retains specificity and is very straightforward. An example of double antibody immunoreactivity detection with two QD-conjugated secondary antibodies is shown in **Fig. 2**.

1. Begin immunostaining as described in **Subheading 3.2., steps 1–4**.
2. Dilute multiple primary antibodies directed toward antigens of interest in the same antibody cocktail. For example, one can make a cocktail containing:
   a. 1 µL Mouse-derived anti-MAP2 antibody.
   b. 1 µL Rabbit-derived anti-GFAP antiserum.
   c. 198 µL PBS-BB.

   For a 1:200 dilution of both primary antibodies.

3. Apply primary antibody cocktail to each slide and incubate overnight at 4°C.
4. Wash the slides three times in PBS for 5 min each. Remove Pap pen from slides (*see* **Note 6**).
5. Prepare a single solution of QD-conjugated secondary antibodies diluted in QD buffer at their respective optimal concentrations. For example, one can make the following secondary antibody cocktail to detect the primary antibodies used in **step 2**:

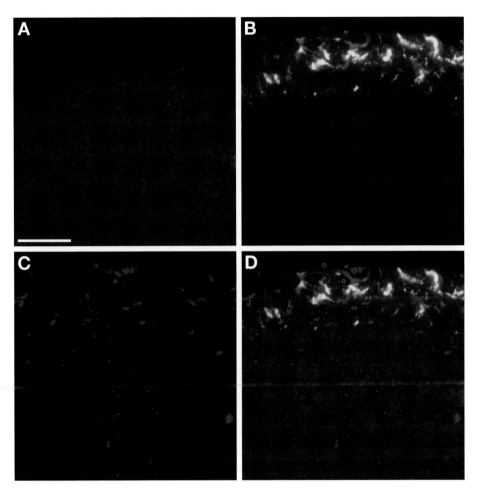

Fig. 2. Simultaneous detection of two antibodies, raised in different species, with two quantum dot (QD)-conjugated secondary antibodies. (**A–D**) Immunostained section of postnatal day 7 mouse brainstem. (**A**) Microtubule-associated protein-2 is immunostained with a mouse monoclonal antibody, subsequently detected with QD 605-conjugated goat anti-mouse secondary antibody, and imaged with a 545 ± 30-nm excitation filter and a 610 ± 75-nm emission filter. (**B**) Glial fibrillary-associated protein (GFAP) is immunostained with a rabbit polyclonal antiserum, subsequently detected with QD 565-conjugated goat anti-rabbit secondary antibody, and imaged with a 470 ± 40-nm excitation filter and a 535 ± 40-nm emission filter. (**C**) Hoescht 33,258 counterstaining of nuclei is imaged with a 360 ± 40-nm excitation filter and a 450 ± 50-nm emission filter. (**D**) Merged image of MAP2, GFAP, and Hoescht 33,258 counterstaining. Scale bar = 40 μm.

    a. 1 µL of QD 605 conjugated goat-derived anti-mouse antibody.
    b. 1 µL of QD 565 conjugated goat-derived anti-rabbit antibody.
    c. 98 µL of QD buffer.

    For a 1:100 dilution of both secondary antibodies.

6. Apply secondary antibody cocktail to each slide and incubate for 1 h at room temperature.
7. Wash the slides three times in PBS for 5 min each. If nuclear counterstaining is desired, dilute stock bisbenzimide (Hoescht 33,258) solution in the PBS during the second wash.
8. Cover slip slides in 90% glycerol in PBS (*see* **Note 10**).
9. View slides with a fluorescence microscope equipped with a 360 ± 40-nm excitation filter and a 400-nm long pass emission filter, which will display both labeled colors. Alternatively, use a narrow-band filter set for each of the QDs used and view the signals individually (*see* **Subheading 3.7.** for information on optimal filter set selection).

3.4. Detection of Single Antibody Immunoreactivity by TSA and QD-Conjugated Streptavidin

QD streptavidin conjugates can be successfully incorporated in enzyme-based signal amplification techniques, including TSA *(4)*. Enzyme-based signal amplification in immunohistochemical detection methods effectively increases the number of fluorescent molecules at antigenic sites of interest *(1)*. Although several enzyme-based signal amplification methods are available, TSA is the most versatile. TSA can greatly increase the sensitivity of immunohistochemical detection, thereby aiding the detection of antigens that are expressed at low levels *(8,9)*. Furthermore, signal amplification can increase the ability to detect antigens within an autofluorescent or otherwise noisy sample. TSA is based on HRP-catalyzed deposition of tyramide-conjugated molecules at sites of antibody binding *(8,10,11)*. One can either deposit fluorophore-conjugated tyramide (direct TSA) or biotinylated tyramide with subsequent detection with a fluorophore-labeled streptavidin (indirect TSA). Next, we outline a protocol in which HRP-conjugated secondary antibody catalyzes the deposition of biotinylated tyramide at sites of primary antibody binding. Deposited biotin is then detected with QD-conjugated streptavidin. **Figure 3** depicts binding of two primary antibodies raised in different species, where one primary antibody is detected by TSA and QD-conjugated streptavidin, and the other primary antibody is detected by QD-conjugated secondary antibody.

1. Follow **Subheading 3.1.** to prepare slides (*see* **Note 5**).
2. Incubate sections in 3% hydrogen peroxide in PBS for 5 min at room temperature. Hydrogen peroxide will help destroy endogenous peroxidase activity in the tissue that could otherwise lead to false-positive signals during the enzyme amplification.

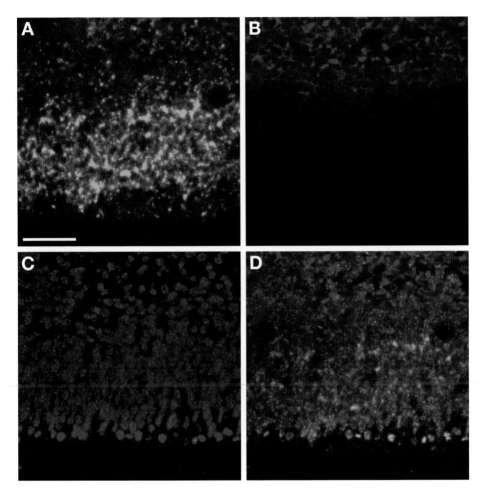

Fig. 3. Sequential detection of two antibodies, raised in different species, with tyramide signal amplification (TSA) and quantum dot (QD)-conjugated streptavidin and with QD-conjugated secondary antibody. **(A–D)** Immunostained section of embryonic day 13 mouse telencephalon. **(A)** Musashi, a marker of neural precursor cells in the ventricular zone, is immunostained with a rabbit polyclonal antiserum, subsequently detected with TSA and QD 525-conjugated streptavidin, and imaged with a 470 ± 40-nm excitation filter and a 535 ± 40-nm emission filter. **(B)** TuJ1, a marker of immature neurons in the intermediate zone, is immunostained with a mouse monoclonal antibody, subsequently detected with QD 605-conjugated goat anti-mouse secondary antibody, and imaged with a 545 ± 30-nm excitation filter and a 610 ± 75-nm emission filter. **(C)** Hoescht 33,258 counterstaining of nuclei is imaged with a 360 ± 40-nm excitation filter and a 450 ± 50-nm emission filter. **(D)** Musashi, TuJ1, and Hoescht 33,258 counterstaining is simultaneously visualized with a 325 ± 25-nm excitation filter and a 400LP emission filter. Scale bar = 40 µm.

3. Wash slides in running tap water for 5 min.
4. Wash sections in PBS for 5 min at room temperature.
5. For primary antibody application, follow **Subheading 3.2., steps 2–7**. Note that when using TSA, the optimal primary antibody concentration is often dramatically reduced compared with non-TSA immunohistochemistry detection techniques.
6. Dilute HRP-conjugated secondary antibody in PBS-BB. Apply 50–200 µL of diluted HRP-conjugated secondary antibody to each slide and incubate for 1 h at room temperature.
7. Wash the slides three times in PBS for 5 min each.
8. Dilute biotinylated tyramide in amplification buffer as recommended by the manufacturer. Apply biotinylated tyramide to each slide.
9. Gently rotate slides in the humidified chamber on an elliptical rotator (25 rpm) for 3–10 min at room temperature.
10. Wash the slides three times in PBS for 5 min each. Remove Pap pen from slides (*see* **Note 6**).
11. Dilute QD streptavidin conjugate in QD buffer at a 1:400 dilution and incubate for 30 min at room temperature prior to use.
12. Apply 50–200 µL of diluted QD streptavidin conjugate to each slide and incubate for 30 min at room temperature.
13. Wash the slides three times in PBS for 5 min each. If nuclear counterstaining is desired, dilute stock bisbenzimide (Hoescht 33,258) solution in the PBS during the second wash.
14. Cover slip slides in 90% glycerol in PBS.
15. View slides with a fluorescence microscope equipped with a 360 ± 40-nm (DAPI) excitation filter and a 400-nm long pass emission filter (*see* **Note 11**) (*see* **Subheading 3.7.** for information on optimal filter set selection.

3.5. Sequential Detection of Multiple Antibody Immunoreactivity by TSA and QD-Conjugated Streptavidin and QD-Conjugated Secondary Antibodies

The protocol described in **Subheading 3.3.** can only be used if the primary antibodies employed are raised in different species. If the antibodies are raised in the same species (e.g., the detection of two mouse-derived primary antibodies), special techniques can be used to retain specificity while using organic fluorophores or QDs *(11)*. Here we will describe a technique called "dilutional neglect" using TSA and multiple QDs. The first primary antibody is detected using TSA and QD-conjugated streptavidin, and the second primary antibody is detected by QD-conjugated secondary antibody. Because TSA significantly increases sensitivity, lower concentrations of primary antibody are required for successful immunostaining. As long as the first antibody that is to be detected with TSA and QD-conjugated streptavidin is used at a lower concentration than that required for detection by QD-conjugated secondary antibodies, the detection method for the second primary antibody will "neglect" the first primary

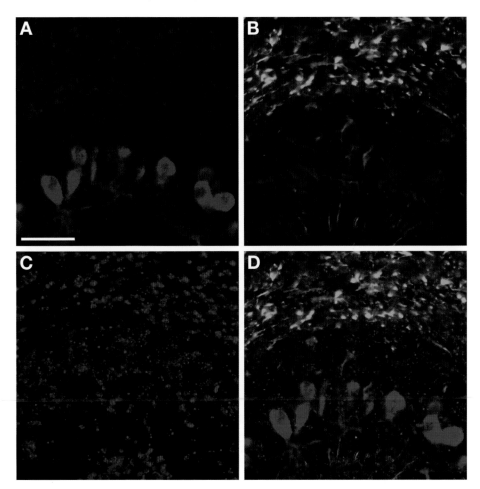

Fig. 4. Sequential detection of two antibodies, raised in the same species, with tyramide signal amplification (TSA) and quantum dot (QD)-conjugated streptavidin and with QD-conjugated secondary antibody. **(A–D)** Immunostained section of postnatal day 7 mouse cerebellum. **(A)** Calbindin is immunostained with a rabbit polyclonal antiserum, subsequently detected with TSA and QD 605-conjugated streptavidin, and imaged with a 545 ± 30-nm excitation filter and a 610 ± 75-nm emission filter. **(B)** Glial fibrillary-associated protein (GFAP) is immunostained with a rabbit polyclonal antiserum, subsequently detected with QD 565-conjugated goat anti-rabbit secondary antibody, and imaged with a 470 ± 40-nm excitation filter and a 535 ± 40-nm emission filter. **(C)** Hoescht 33,258 counterstaining of nuclei is imaged with a 360 ± 40-nm excitation filter and a 450 ± 50-nm emission filter. **(D)** Calbindin, GFAP, and Hoescht 33,258 counterstaining is simultaneously visualized with a 325 ± 25-nm excitation filter and a 400LP emission filter. Scale bar = 40 μm.

antibody and only indicate binding of the second primary antibody. QDs can easily be integrated into this type of protocol, as outlined next. In **Fig. 4**, TSA using QD-conjugated streptavidin is combined with QD secondary antibody to perform double labeling where the two primary antibodies used are raised in the same species.

1. Prepare slides as described in **Subheading 3.1.** (*see* **Note 5**).
2. Begin immunostaining as described in **Subheading 3.4., steps 2–4**. Next, complete **Subheading 3.2., steps 2–4**.
3. Prepare the first primary antibody diluted in PBS-BB. For example, we will use rabbit derived anti-calbindin antiserum at a 1:10,000 dilution. At this dilution, calbindin antiserum binding will not be detected by QD-conjugated secondary antibody, which requires a 1:100 dilution. Apply primary antibody to each slide and incubate overnight at 4°C. *Do not* apply the second primary antibody at this time.
4. Wash the slides three times in PBS for 5 min each.
5. Dilute HRP-conjugated secondary antibody in PBS-BB. Apply 50–200 µL of diluted HRP-conjugated secondary antibody to each slide and incubate for 1 h at room temperature.
6. Wash the slides three times in PBS for 5 min each.
7. Dilute biotinylated tyramide in amplification buffer as recommended by the manufacturer. Apply biotinylated tyramide to each slide.
8. Gently rotate slides in the humidified chamber on an elliptical rotator (25 rpm) for 10 min at room temperature.
9. Wash the slides three times in PBS for 5 min each.
10. Repeat **Subheading 3.2., steps 2–4**.
11. Prepare the second primary antibody diluted in PBS-BB. In our example, we will use rabbit derived anti-GFAP antiserum at a 1:100 dilution. Apply primary antibody to each slide and incubate overnight at 4°C.
12. Wash the slides three times in PBS for 5 min each.
13. Prepare QD-conjugated secondary antibody and QD streptavidin conjugate in a reagent cocktail at their respective concentrations. Be sure the QDs used emit fluorescence at different wavelengths. For example, the following will detect the primary antibodies used in **steps 3** and **11**:
    a. 1 µL of QD 605 streptavidin conjugate.
    b. 4 µL of QD 565-conjugated goat-derived anti-rabbit antibody.
    c. 395 µL of QD buffer.

    For a 1:400 dilution of QD 605-conjugated streptavidin and a 1:100 dilution of QD 565-conjugated goat derived anti-rabbit antibody.

14. Apply reagent cocktail to each slide and incubate for 1 h at room temperature.
15. Wash the slides three times in PBS for 5 min each. If nuclear counterstaining is desired, dilute stock bisbenzimide (Hoescht 33,258) solution in the PBS during the second wash.
16. Cover slip slides in 90% glycerol in PBS (*see* **Note 10**).

17. View slides with a fluorescence microscope equipped with a 360 ± 40-nm excitation filter and a 400-nm long pass emission filter, which will display both labeled colors. Alternatively, use a narrow-band filter set for each of the QDs used and view the signals individually (*see* **Subheading 3.7.** for information on optimal filter set selection).

3.6. Triple Labeling Using QD-Conjugated Streptavidin and Multiple QD-Conjugated Secondary Antibodies

In practice, several QD-conjugated secondary antibodies may be used to simultaneously and specifically detect multiple primary antibody binding in a given tissue sample, provided that all primary antibodies are raised in different species. Therefore, the protocols in **Subheadings 3.3.** and **3.5.** can be combined to achieve triple labeling using QDs. Several primary antibodies, which will be detected using different QD-conjugated secondary antibodies, should be prepared as an antibody cocktail and applied to slides. If enzyme-based signal amplification is desired, it may be performed for one antibody before the use of an antibody cocktail. An example of triple primary antibody immunoreactivity detection with a combination of the techniques described here is depicted in **Fig. 5**.

3.7. Selecting Filter Sets for Use With QDs and Immunohistochemical Detection

To successfully view QDs in immunohistochemical applications, the appropriate filter sets must be used. Investigators may misinterpret their results if an incorrect or suboptimal filter set is used to view the staining. General red/green/blue emission filters may be adequate to passage the narrow emission wavelengths of QDs. It is recommended to use custom filter sets for QD viewing, but purchasing and maintaining these filters in addition to standard filter sets may not be cost effective. Based on our observations, we discuss next several aspects of filter set selection for immunohistochemical applications.

1. Quantum Dot Corporation recommends a 460-nm short-pass excitation filter, a 475-nm dichroic beamsplitter, and a narrow-band pass filter centered at the wavelength corresponding to the QD to be viewed. For example, a QD 605-conjugated secondary antibody should be viewed with a 605 ± 20-nm narrow-band pass emission filter. Custom filtersets that include these filters are currently available from Chroma Technology or Omega Optical.
2. QDs are optimally excited at wavelengths below 460 nm, and the use of a short pass 460-nm filter will provide maximal excitation. No currently available QDs emit light at or below 460 nm. However, several other dyes, such as DAPI and Hoescht 33,258, emit fluorescence at or around 460 nm. If it is desired to use these dyes concurrently with QDs, an ultraviolet excitation filter that does not exceed

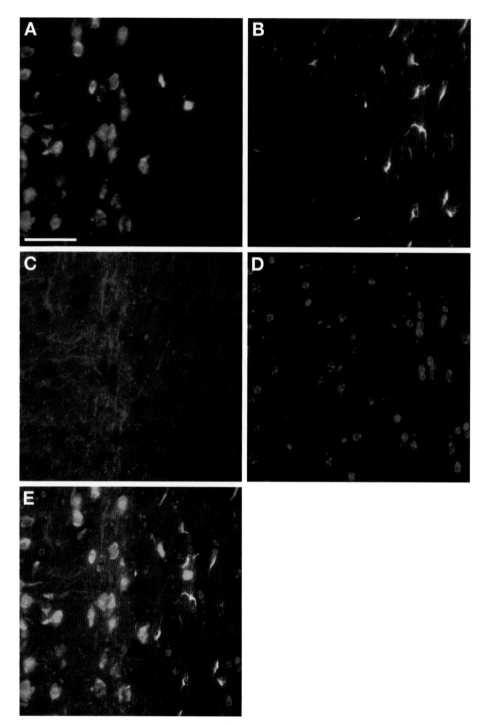

400 nm can be used. We have had good results using a 325 ± 25-nm or 360 ± 40-nm excitation filter with QDs. Using a 460-nm short pass excitation filter would preclude the simultaneous visualization of QDs and fluorescent dyes such as DAPI or Hoescht 33,258.

3. Conventional narrow-band filters for fluorophores, such as Cy3 or FITC, can be used to visualize certain QDs. For example, a Cy3 filter set containing a 545 ± 30-nm excitation filter, a 570-nm long pass dichroic beamsplitter, and a 610 ± 75-nm emission filter would adequately allow the visualization of QD 605. The 610 ± 75-nm emission filter, while not centered on the 605-nm wavelength, would allow the passage of emitted light at 605 nm because of its relatively wide bandwidth (75 nm). Furthermore, a FITC filter set containing a 470 ± 40-nm excitation filter, a 500-nm long pass dichroic beamsplitter, and a 525 ± 40-nm emission filter would transmit fluorescence from both QD 525 and 565. Therefore, an investment in custom filter sets for QD immunohistochemistry is not absolutely required. However, specific visualization of QD 565 would require a custom filterset because that wavelength does not correspond to any other commonly used fluorophore. The custom filtersets are optimized for detection of QDs and can be expected to produce the best signal-to-noise ratio.

4. If the simultaneous visualization of several QDs is desired, a long-pass emission filter can be used. For example, QD 525, 565, and 605 can be simultaneously visualized with a 470 ± 40-nm excitation filter, a 510-nm long-pass dichroic beamsplitter, and a 520-nm long-pass emission filter. In this way, all three QDs can be seen at the same time, and currently our laboratory uses a similar filter for QD visualization. If a fourth fluorescent dye is to be visualized (such as DAPI or Hoescht 33,258), one can use a 325 ± 35-nm excitation filter, a 370-nm long-pass dichroic beamsplitter, and a 400-nm long-pass emission filter. When using long-pass emission filters, there may be a decrease in the signal-to-noise ratio because

---

Fig. 5. Sequential detection of three antibodies with tyramide signal amplification (TSA) and quantum dot (QD)-conjugated streptavidin and with two QD-conjugated secondary antibodies. **(A–E)** Immunostained section of adult mouse brain. **(A)** NeuN, a marker of mature neurons, is immunostained with a mouse monoclonal antibody, subsequently detected with TSA and QD 525-conjugated streptavidin, and imaged with a 470 ± 40-nm excitation filter and a 535 ± 40-nm emission filter. **(B)** Glial fibrillary-associated protein (GFAP) is immunostained with a rabbit polyclonal antiserum, subsequently detected with QD 565-conjugated donkey anti-rabbit secondary antibody, and imaged with a 470 ± 40-nm excitation filter and a 535 ± 40-nm emission filter. **(C)** Microtubule-associated protein-2 (MAP2) is immunostained with a mouse monoclonal antibody, subsequently detected with QD 605-conjugated goat anti-mouse secondary antibody, and imaged with a 545 ± 30-nm excitation filter and a 610 ± 75-nm emission filter. **(D)** Hoescht 33,258 counterstaining of nuclei is imaged with a 360 ± 40-nm excitation filter and a 450 ± 50-nm emission filter. **(E)** Merged image of NeuN, GFAP, MAP2, and Hoescht 33,258 counterstaining. Scale bar = 40 μm.

of the additional wavelengths of light transmitted, thereby leading to increased visible autofluorescence.

4. Notes

1. The protocols outlined here are robust in our laboratory and should be used as starting points for the development of protocols in other laboratories. There are a large number of potential fixatives, and we recommend that several fixatives be tried during protocol optimization. The particular fixative used is selected based on the tissue to be immunostained, the requirement for perfusion, the stability or availability of the antigen to be detected, and the characteristics of the antibody to be used. Furthermore, there are also many methods of performing antigen retrieval, and we recommend that several be tried for each antigen and tissue.

2. Washing slides can best be accomplished by placing slides in a suitable glass slide rack that will fit inside a rectangular staining dish and transferring the entire rack from one wash solution to another. Transferring individual slides between various solutions for washing is typically not necessary. Also, slides can be washed by removing the entire slide rack from the staining dish, quickly emptying the dish into the sink, replacing the rack in the dish, and rapidly pouring a new solution in the dish. Care should be taken to minimize the time slides are exposed to air.

3. We recommend establishing a "deparrafinization station" to be used by several investigators in a laboratory. Use two staining dishes that contain CitriSolv and three that contain isopropanol. Use each dish sequentially to perform deparrafinization and rehydration. After several rounds of use, replace dish 1 with dish 2, and dish 2 with dish 3. Then, clean dish 1, fill with solution, and use as dish 2 (CitriSolv) or 3 (isopropanol).

4. At no time should the tissue on a slide be allowed to dry. Dried tissue is perhaps the most frequent cause of suboptimal or uninterpretable staining. The tissue must remain moist at all times. To ensure that slides remain wet, they must be quickly transferred between containers during washes. Also, the slides should be inspected periodically during incubations with antibody or detection reagents to ensure that solution adequately covers the tissue. Even partial or temporary dryness may lead to nonspecific absorption of reagents to the tissue resulting in high background and/or nonspecific signals.

5. When performing any type of immunohistochemistry, control slides should be run in parallel that do not receive primary antibody but do receive all subsequent detection reagents. These negative control slides will illustrate any background or false-positive signals generated by the detection method that is independent of primary antibody binding. Some tissue has significant autofluorescent material that will be evident in these negative control slides. In addition, we have found that if used at inappropriately high concentrations, QD reagents may nonspecifically label neurons *(4)*. This nonspecific labeling will also be evident in negative control slides, and will allow for better interpretation of truly labeled slides. Nonimmune serum controls or preabsorbed primary antibody controls are also useful for interpreting immunohistochemical results.

6. Pap pens are very useful to create a boundary on a slide to prevent the lateral movement of solution from capillary action. This way, a smaller volume of antibody/ reagent can be used to cover tissue completely. However, Pap pen may interfere with QD signals and photostability *(7)*, and we recommend removing Pap pen prior to the addition of QD reagents. Removal is accomplished by using a damp paper towel or a fingernail. Pap pen is not absolutely required and a grease pencil, scoring with a diamond tip, or an aerosolized hydrocarbon coating around the tissue can alternatively be used with similar results. Quantum Dot Corporation suggests the use of the ImmEdge Pap pen from Vector Labs to minimize interference with signal intensity.
7. To remove antibody or reagent solutions from a slide, firmly tap the slide on its side onto the lab bench. Then, use a folded dry paper towel to dab over the Pap pen boundary. There should be no wetness "over" the boundary. Otherwise, solution that is added within the boundary will easily pass to the other side via capillary action. This process must be done quickly and carefully to ensure the tissue section does not dry or become scratched or damaged.
8. While slides are in the humidified chamber (slide folder), ensure that the folder remains as horizontal as possible. Be sure that any refrigerator shelves that are used to house slides overnight are level, so that slides remain adequately wet. Also, when removing slides from the chamber for washing, the underside of the slide folder "leaves" should be inspected and washed if any residual antibody or reagent is present. The slide folder can then be rinsed and reused for the next step of immunostaining.
9. When washing slides, it is important to use large volumes of fresh PBS. One can also agitate slides at 100 rpm on a rotator for better washing.
10. Slides that are cover slipped with 90% glycerol in PBS may have undesirable air bubbles beneath the cover slip. To overcome this problem, the cover slip should be gently removed and reapplied until the air bubbles are either eliminated or are not over any pieces of tissue.
11. QD fluorescence may require a brief period of illumination prior to viewing to maximize their signal-to-noise ratio. The immunostained slide should be placed in the light path of the microscope and for 2–4 min for the specific fluorescence to become optimally apparent.

Acknowledgments

We would like to thank all the members of the Roth laboratory for their insights and assistance. These protocols were developed in part by support from NIH grants NS35107 and NS41962. RSA received support from the UAB Medical Scientist Training Program (NIH grant 08361).

References

1. Roth, K. A. and Baskin, D. G. (2005) Enzyme-based fluorescence amplification for immunohistochemistry and *in situ* hybridization. In: *Molecular Morphology in Human Tissues: Techniques and Applications,* (Tubbs, R. R. and Hacker, G. W., eds.), CRC Press, Washington, D.C., pp. 65–80.

2. Bruchez, M., Jr., Moronne, M., Gin, P., Weiss, S., and Alivisatos, A. P. (1998) Semiconductor nanocrystals as fluorescent biological labels. *Science* **281,** 2013–2016.

3. Wu, X., Liu, H., Liu, J., et al. (2003) Immunofluorescent labeling of cancer marker Her2 and other cellular targets with semiconductor quantum dots. *Nat. Biotechnol.* **21,** 41–46.

4. Ness, J. M., Akhtar, R. S., Latham, C. B., and Roth, K. A. (2003) Combined tyramide signal amplification and quantum dots for sensitive and photostable immuno-fluorescence detection. *J. Histochem. Cytochem.* **51,** 981–987.

5. Watson, A., Wu, X., and Bruchez, M. (2003) Lighting up cells with quantum dots. *BioTechniques* **34,** 296–303.

6. Jaiswal, J. K., Mattoussi, H., Mauro, J. M., and Simon, S. M. (2003) Long-term multiple color imaging of live cells using quantum dot bioconjugates. *Nat. Biotechnol.* **21,** 47–51.

7. Qdot Streptavidin Conjugates User Manual. (2005) Molecular Probes, Inc., Eugene, OR, pp. 1–14. http://probes.invitrogen.com/media/pis/mp19000.pdf.

8. Bobrow, M. N., Litt, G. J., Shaughnessy, K. J., Mayer, P. C., and Conlon, J. (1992) The use of catalyzed reporter deposition as a means of signal amplification in a variety of formats. *J. Immunol. Methods* **150,** 145–149.

9. van Gijlswijk, R. P., Zijlmans, H. J., Wiegant, J., et al. (1997) Fluorochrome-labeled tyramides: use in immunocytochemistry and fluorescence in situ hybridiza-tion. *J. Histochem. Cytochem.* **45,** 375–382.

10. Bobrow, M. N., Harris, T. D., Shaughnessy, K. J., and Litt, G. J. (1989) Catalyzed reporter deposition, a novel method of signal amplification. Application to immunoassays. *J. Immunol. Methods* **125,** 279–285.

11. Shindler, K. S. and Roth, K. A. (1996) Double immunofluorescent staining using two unconjugated primary antisera raised in the same species. *J. Histochem. Cytochem.* **44,** 1331–1335.

3

Quantum Dots for Multicolor Tumor Pathology and Multispectral Imaging

Johbu Itoh and Robert Yoshiyuki Osamura

Summary

Quantum dots (QDs) are new nanocrystal semiconductor fluorophores consisting of a cadmium selenide core and zinc sulfide or cadmium sulfide shell. They have many advantages over conventional fluorophores including prolonged signal because of photostability, and reinforcement of weak positive reactions. In this study, we have used QDs to intensify fluorescent signals in immunohistochemistry for pathology diagnostics. In addition, we introduce a new confocal laser scanning microscopy analysis method called the META system (Carl Zeiss, Jena, Germany) in which a dye spectrum is utilized. This system ensures optimum specimen illumination and efficient collection of reflected or emitted light and uses an innovative way of separating fluorescent emissions. Our results suggest that very weak immunoreactions seen by traditional immunohistochemical techniques can be greatly intensified, a useful feature for pathology diagnostics.

Key Words: Quantum dot; fluorescent; spectrum analysis; confocal laser scanning microscopy; pathology diagnosis.

1. Introduction

Evaluation of subcellular structures by light microscopy is important in many fields including pathology diagnostics. In addition, immunohistochemistry, immunocytochemistry, and enzyme histochemistry are key methods used to localize particular substances in subcellular organelles *(1–4)*.

The immunofluorescent antibody method is the standard technique for the detection of antigens and proteins in cells and tissue. However, conventional fluorophores fade and production of a permanent specimen is difficult. In addition, fluorescence-dependent imaging is limited in spatial resolution by the wavelength of light, and is generally not suitable for electron microscopy. However, a few studies have been published demonstrating electron microscopy

From: *Methods in Molecular Biology, vol. 374: Quantum Dots: Applications in Biology*
Edited by: M. P. Bruchez and C. Z. Hotz © Humana Press Inc., Totowa, NJ

observation of immunofluorescent signals *(5,6)*. Furthermore, the use of transmission electron microscopy in conjunction with immunolabeling such as
horseradish peroxidase (HRP)-diaminobenzidine (DAB) conjugated osmium
acid has proved to be advantageous for high-resolution structural studies *(7,8)*.

Quantum dots (QDs), a new type of fluorescent probe, have several advantages over conventional fluorophores including photostability, which prolongs
the signal, and intense fluorescence, which reinforces weak immunological
interactions *(9–15)*. In addition, the electron density and elemental composition
of QDs permit the extension of their use to immunoelectronmicroscopy *(16)*.

QDs, which are commercially available as QD conjugates from Quantum
Dot Corporation (Hayward, CA), consist of a semiconductor core of cadmium
selenide coated with a shell of zinc sulfide. An additional polymer layer
enhances water solubility and enables conjugation to streptavidin or other
biomolecules such as IgG *(17)*. These particles have the optical properties of
high brightness, photostability, narrow emission spectra, and an apparently
large Stokes' shift *(18)*. Slight changes in the size of the semiconductor core
change the emission spectrum, resulting in a redder color as the size of the core
increases. Many studies have utilized these features to obtain good imaging
results *(9–13,19–21)*. In this study we focus on two advantages of QDs:
improvement in sensitivity, and the use of a single light source to excite different wavelengths.

Confocal laser scanning microscopy (CLSM) is another technique that has
been used for detection of fluorescent signals *(22–24)*, and recently a new system
of CLSM spectrum analysis, the META system, has been introduced to the field
of cell biology *(25,26)*. The unique scanning module is the core of the LSM 510
META. It contains motorized collimators, scanning mirrors, individually
adjustable and positionable pinholes, and highly sensitive detectors including
the META detector. These components are arranged to ensure optimum specimen
illumination and efficient collection of reflected or emitted light. A highly efficient
optical grating provides an innovative way of separating the fluorescence emissions in the META detector. The grating projects the entire fluorescence spectrum
onto the 32 channels of the META detector to acquire the spectral signature for
each pixel of the scanned image. Subsequently, the component dyes can be
digitally separated.

First, a Lambda Stack is acquired to record the spectral signature of the
specimen. The Lambda Stack records spectral distributions of fluorescence
emissions in 10-nm steps. As these spectrally resolved images are recorded
simultaneously, this step is completed in minimum time, which not only is
friendly to delicate specimens but also reliably captures fast dynamic processes.
Here, we describe the application of a QD fluorophore to the detection of intracellular antigens in cells and/or tissue sections by CLSM-META analysis.

2. Materials

1. Human pituitary adenoma tissue: surgically resected pituitary adenoma tissues were fixed in cold 4% paraformaldehyde and processed in paraffin wax. The paraffin sections (3–7 μm) were then prepared for immunohistochemistry.
2. Human breast cancer tissue: surgically resected breast cancer tissues were fixed in 10% formaldehyde and processed in paraffin wax. Thick sections (5–7 μm) were prepared using a microtome and submitted to subsequent double- and/or triple-labeling immunohistochemistry.
3. The primary antibodies used to detect adrenocorticotropic hormone (ACTH) and growth hormone (GH) were both rabbit polyclonal antibodies (1:200, DAKO, Santa Barbara, CA).
4. The secondary antibodies used for adenoma labeling were either Alexafluor 488 conjugated to goat anti-rabbit IgG (1:10; Molecular Probes, Inc., Eugene, OR), or streptavidin-labeled QD605 (1:100; Quantum Dot Corporation).
5. DAB solution: 0.05 M Tris-HCl buffer (pH 7.6) containing 30 mg/100 mL DAB, 65 mg/mL sodium azide, and 0.003% hydrogen peroxide.
6. Human epidermal growth factor receptor (HER)-2 expression was tested using the Hercep Test Kit (DAKO, Carpinteria, CA, cat. no. K-5204) according to the manufacturer's instructions.
7. Estrogen receptor (ER) express (Envision System, K4006, DAKO), an enzyme polymer-enhanced method kit, was used.
8. Epidermal growth factor receptor (EGFR) expression was tested using the EGFR pharmDx™ Kit (DAKO, cat. no. K1492) according to the manufacturer's instructions.
9. The secondary antibodies for breast labeling were as follows: anti-rabbit-IgG-QD605, anti-mouse-IgG-QD565, streptavidin-labeled QD605 (1:100; Quantum Dot Corp.), HRP-conjugated sheep anti-rabbit IgG (1:50; Amersham, Poole, UK), HRP-conjugated sheep anti-mouse IgG (1:50; Amersham), and alkaline phosphate (ALP)-conjugated sheep anti-rabbit IgG (1:50; DAKO).
10. The peroxidase coloring reaction for breast labeling was performed with 3-3′ DAB (Dojindo Co. Ltd., Kumamoto, Japan). Fast Red staining buffer (DAKO, cat. no. K0597) containing 5 M levamisole was prepared immediately before application to tissue sections.
11. SlowFade® DABCO-containing mounting medium (Invitrogen Corp., Carlsbad, CA).
12. Blocking buffer: 0.05% Triton X-100, 2% bovine serum albumin (Sigma-Aldrich, St. Louis, MO) in phosphate-buffered saline (PBS).
13. Secondary antibodies: the secondary antibodies were as follows: anti-rabbit-IgG-QD605, anti-mouse-IgG-QD565, streptavidin-labeled QD605 (1:100; Quantum Dot Corp.), HRP-conjugated sheep anti-rabbit IgG (1:50; Amersham), HRP-conjugated sheep anti-mouse IgG (1:50; Amersham), and ALP-conjugated sheep anti-rabbit IgG (1:50; DAKO). The peroxidase coloring reaction was performed with 3-3′ DAB (Dojindo Co. Ltd.). Fast Red staining buffer (DAKO, cat. no. K0597) containing 5 M levamisole was prepared immediately before application to tissue sections and color was developed for 4–10 min.

3. Methods
3.1. Preparation of Paraffin-Embedded Tissue Sections

1. Specimens from surgically resected tissues were cut into 3- to 4-mm blocks and fixed.
2. The tissues were then dehydrated and cleared in an alcohol/xylene series, followed by infiltration with melted paraffin not exceeding 60°C. Properly fixed and embedded tissue blocks expressing ACTH and GH proteins will keep indefinitely prior to sectioning and slide mounting if stored in a cool place (15–25°C).
3. Tissue specimens were then cut into 3- to 5-μm sections, positioned on poly-L-lysine-coated slides, and placed in drying racks.
4. Excess water was removed by placing the slide racks on an absorbent towel and drying at room temperature for 1 h. The racks were then incubated at 56–60°C for 1 h. This process was repeated to remove any remaining water.
5. Once the slides were dry, they were cooled at room temperature until the paraffin hardened.
6. Tissue sections were then mounted on 3-amino-propylmethoxysilane-coated slides to preserve antigenicity.
7. After immunostaining, the slides were mounted in a solution containing 50 mM Tris-HCl (pH 8.0), 0.2 M 1,4-diazabicyclo-2,2,2-octane (DABCO; Wako), and 90% glycerol.

3.2. Immunohistochemistry
3.2.1. Human Pituitary Adenoma

1. Immunohistochemical analysis of hormone-secreting endothelial cells was performed by indirect immunofluorescent and immunoperoxidase-labeled antibody methods. The procedures were performed as described by Nakane *(27)*.
2. The primary antibodies used to detect ACTH and GH were both rabbit polyclonal antibodies (1:200; DAKO).
3. The secondary antibodies used were either Alexafluor 488 conjugated to goat anti-rabbit IgG (1:10; Molecular Probes, Inc.), or streptavidin-labeled QD605 (1:100; Quantum Dot Corp.).
4. Slides requiring binding to an avidin–biotin complex were incubated with this solution for 40 min at room temperature and then washed in distilled water.
5. Nuclear counterstaining was performed using methyl green solution.
6. Enzyme immunohistochemical visualization was performed by immersing the sections in 0.05 M Tris-HCl buffer (pH 7.6) containing 30 mg/dL DAB, 65 mg/dL sodium azide, and 0.003% hydrogen peroxide.

3.2.2. Human Breast Cancer
3.2.2.1. ANTIGEN RETRIEVAL METHODS

1. HER 2: the antigen retrieval step was performed at 99°C for 40 min in citrate buffer (pH 6.0) (10 mM citric acid).

2. ER clone ID5, DAKO: the antigen retrieval step was performed by incubation at 99°C for 40 min in DAKO target retrieval solution high pH (DAKO, cat. no. S3307).
3. EGFR: the antigen retrieval step was performed by incubation for 5 min at room temperature in a protease K solution (pH 6.5).

3.2.2.2. IMMUNOSTAINING PROCEDURE: HER2

1. HER2 expression was tested using the Hercep Test Kit (DAKO, cat. no. K-5204) according to the manufacturer's instructions.
2. The slides were dewaxed in xylene baths for 5 min each and rehydrated in an alcohol series (95 and 70%) for 3 min in each bath.
3. Subsequent to antigen retrieval slides were cooled at room temperature for 20 min, and the peroxidase activity was blocked with 3% hydrogen peroxide in methanol for 5 min.
4. The slides were then rinsed in distilled water.
5. The specimens were then incubated with the appropriate primary antibody for 30 min at room temperature in a humidified chamber.
6. After rinsing with PBS (pH 7.6) at room temperature, the slides were incubated with the secondary antibody (anti-rabbit-HRP [1:100] or anti-rabbit-QD605 [1:100]) for 1 h at room temperature, followed by three rinses for 5 min each in PBS.
7. Following incubation in DAB for 5 min at room temperature and rinsing in distilled water, the slides were counterstained with methyl green for 1 min, dehydrated in an alcohol series, and mounted in a solution containing 50 mM Tris-HCl buffer and DABCO (Wako).

3.2.2.3. IMMUNOSTAINING PROCEDURE: ER

1. ER expression was tested using the DAKO ENVSIO kit/ HRP(DAB) (DAKO, cat. no. K4006).
2. Slides were initially dewaxed in xylene baths for 5 min followed by rehydration in an alcohol series (95 and 70%) for 3 min in each bath.
3. After washing in deionized water, the slides were incubated in epitope retrieval solution at 95°C for 40 min and then cooled for 20 min at room temperature.
4. Remaining liquids were removed by wiping around the selected area.
5. The slides were incubated for 5 min with peroxidase blocking reagent and rinsed in deionized water and then Tris-HCl buffer.
6. Primary antibody or negative control reagent was applied to the selected area, and the slides were incubated for 1 h at room temperature.
7. After rinsing with PBS (pH 7.6) at room temperature, the slides were incubated with the secondary antibody for 1 h at room temperature (anti-mouse-HRP;1:100 or anti-mouse-QD565; 1:100), followed by three rinses of 5 min each in PBS.
8. The specimen was then covered with 100 μL of visualization reagent, incubated for 30 min, rinsed, and covered with 100 μL of HRP-DAB for 10 min.

9. After rinsing with deionized water, counterstaining was performed by incubation for 20–30 min in ToTo-3 (2.4 nM in PBS [dimeric cyanine nucleic acid dyes]; Molecular Probes, Inc.) or ToTo-3 in DABCO.
10. The selected area was then covered with a cover slip using SlowFade DABCO-containing mounting medium.
11. The sample was incubated in SlowFade solution for 30 min prior to visualization to improve the signal-to-background ratio.
12. Slides were stored protected from light at room temperature or at 4°C.

3.2.2.4. EGFR

1. EGFR expression was detected using the EGFR pharmDx kit (DAKO, cat. no. K1492) according to the manufacturer's instructions.
2. Following antigen retrieval, peroxidase activity was blocked with 3% hydrogen peroxide in methanol for 5 min, and the slides were rinsed in distilled water.
3. The slides were then incubated with the primary antibody for 30 min at room temperature in a humidified chamber.
4. After rinsing with PBS (pH 7.6) at room temperature, the slides were incubated with the secondary antibody for 1 h at room temperature, followed by three rinses of 5 min each in PBS (anti-rabbit-HRP [1:100] or anti-rabbit-QD605 [1:100]).
5. Following incubation in DAB for 5 min at room temperature and rinsing in distilled water, the slides were counterstained with methyl green for 1 min, dehydrated in an alcohol series, and mounted in a solution containing 50 mM Tris-HCl buffer and DABCO (Wako).

3.2.2.5. QDOT CONJUGATE PROTOCOL

1. A specimen of the breast tumor was selected for tissue processing, and fixed immediately.
2. After treatment, specimens underwent antigen retrieval.
3. Each slide was rinsed three times with PBS for 5 min and then three times with Tris-buffered saline (pH 7.0) for 3 min.
4. Blocking buffer was then added and slides were incubated in a humidified chamber for 30 min at 37°C.
5. Primary antibodies were diluted in blocking buffer, and concentration was optimized for each antibody with a starting range of 1–10 µg antibody/mL. Blocking buffer was then aspirated and the primary antibody solution was added.
6. Slides were incubated in a humidified chamber for 1 h at 37°C or overnight at 4–8°C.
7. Slides were then washed four times with Tris-buffered saline for approx 3 min each wash, and blocking buffer was added to each slide.
8. These slides were then incubated in a humidified chamber for 10 min at 37°C.
9. Secondary antibodies were diluted as Qdot secondary antibody conjugate in blocking buffer to a final concentration of 2–20 nM (1:500–1:50).
10. Blocking buffer was aspirated, Qdot secondary antibody conjugate solution was added, and slides were incubated for 1 h at room temperature in a humidified chamber.

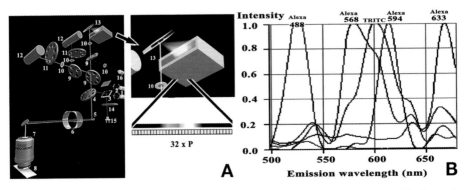

Fig. 1. The META system uses highly sensitive multiple detectors with 32 channels to ensure optimum specimen illumination and efficient collection of reflected or emitted light. A highly efficient optical grating provides an innovative way of separating the fluorescent emissions, allowing a spectral signature to be acquired for each pixel of the scanned image. Spectral signatures of the fluorescent emissions detected in the region of interest are shown. **(A)** 1, optical fibers; 2, motorized collimators; 3, beam combiner; 4, main dichroic beamsplitter; 5, scanning mirrors; 6, scanning lens; 7, objective lens; 8, specimen; 9, secondary dichroic beamsplitters; 10, confocal pinhole; 11, emission filters; 12, photomultiplier; 13, META detector; 14, neutral density filters; 15, monitor diode; 16, fiber out. **(B)** Conventional fluorophore spectrum wavelength.

11. Slides were then washed with cold wash buffer three times for 3 min each. After washing, slides were mounted.

3.3. Spectrum Analysis System

Imaging of specimens was performed using a Carl Zeiss LSM-510 META confocal laser scanning microscope, and the META system was employed for computer-assisted image spectrum analysis (Carl Zeiss). Lambda stack mode and online fingerprinting mode were used. The META system (Carl Zeiss) is a new analysis technique that uses highly sensitive detectors to capture a spectrum of fluorescent signals. It provides optimum specimen illumination, and the detectors have 32 channels to ensure efficient collection of reflected or emitted light. A highly efficient optical grating provides an innovative way of separating the fluorescent emissions. Thus, the spectral signature is acquired for each pixel of the scanned image. This signature can be subsequently used for the digital separation of component dyes. Spectral signatures of the fluorescence emissions detected in the regions of interest are shown in **Fig. 1.**

1. Use of multiple excitation wavelengths. **Figure 2** shows the use of multiple laser excitation wavelengths (Ar458, Ar488, Ar514, and HeNe543nm) to detect the same QD emission wavelength (610 nm). Sensitivity and quantitation are

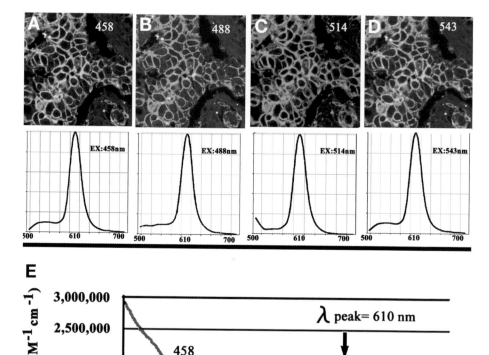

Fig. 2. Multiple laser excitation wavelengths (**A**) Ar458; (**B**) Ar488; (**C**) Ar514; and (**D**) HeNo543nm) were used to detect a single fluorescent emission wavelength. Excitation using two or more lasers for one QD fluorophore is also shown (**E**).

improved over conventional fluorophores because of the photostability of QDs. All fluorescent colors can be excited with one source (one laser or filter).

2. Combined observation of conventional and QD fluorophores. **Figure 3** shows an example of a combined META analysis. This image shows the QD605 META spectral signatures of the fluorescent emissions detected in the regions of interest. The immunohistochemical colocalization by CLSM of GH and ACTH in paraffin sections of a GH-secreting adenocarcinoma (GHoma) of the human pituitary

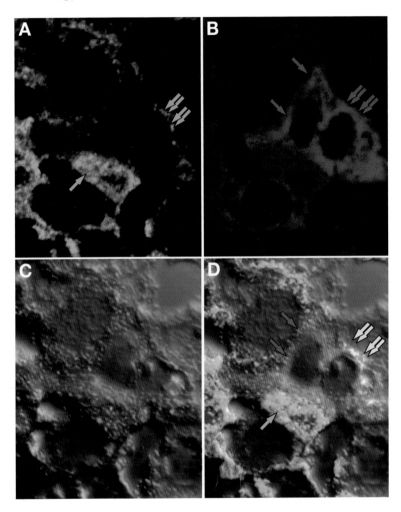

Fig. 3. Immunohistochemical localization of growth hormone (GH) and adrenocorticotropic hormone (ACTH) in human pituitary GH-secreting adenoma. **(A)** GH; **(B)** ACTH; **(C)** transmittance; **(D)** merged image. Simultaneous detection of conventional fluorophore and quantium dots by confocal laser scanning microscopy is shown. ACTH is detected using streptavidin-labeled QD605 (red). GH is identified by Alexa 488 (green). Yellow indicates colocalization of ACTH and GH. The excitation wavelength is 488 nm (argon laser) for both fluorophores.

gland is shown. The conventional fluorophore, Alexa 488 (green) was used to label GH, and ACTH was labeled with streptavidin-conjugated QD605 (red). Yellow indicates colocalization of GH and ACTH. Fluorescence from both Alexa 488 and QD605 were detected simultaneously with a single excitation source, a 488 argon ion laser. Contrary to expectations, the QD signal permeated the sample to a depth of approx 5–6 microns (data not shown), which is convenient for electron-microscopic observation. The cause of this deep permeation is unknown, and is currently under investigation.

3. Combined observation of multiple quantum dot (QD) and conventional fluorophore counterstaining in human breast cancer tissues. **Figure 4** shows an unmixed image of multiple quantum dots probes using META system. The immunohistochemical multiple detection was using QD and conventional counterstaining fluorophores, ToTo-3. Three colors were detected simultaneously: anti-rabbit-IgG-Qd605 (HER2; red), anti-mouse-IgG-Qd565 (ER; green), and ToTo-3 (blue; nuclear counterstaining). Using the unmixed function of LSM510META system, all three colors were individually detected. **Figure 4E** showed individual QD conjugates with emissions in the META detection spectrum patterns. Note: the simultaneous multiple color detections were visualized individually as immunoreactions by CLSM-META analysis.

4. Enzyme-labeled antibodies detected by light microscopy compared with QD probes and CLSM. **Figure 5** shows a comparison of immunohistochemical detection in human breast tissue using enzyme-labeled antibodies and light microscopy (**Fig. 5A–C**), vs QD-labeled antibodies and CLSM with META analysis (**Fig. 5D–E**). **Figure 5A,D** show HER2 expression, **Fig. 5B,E** show EGFR expression, and **Fig. 5C,F** are merged images of **5A,B,D,E**, respectively. Secondary antibodies were conjugated to ALPase-Fast red (red) for HER2 detection in **Fig. 5A,C** and HRP-DAB (green) for EGFR detection in **Fig. 5B,C**. Secondary antibodies used for CLSM-META detection were conjugated to QD605 (red; **Fig. 5D,F**), and QD565 (green; **Fig. 5E,F**). Colocalization of HER2 and EGFR appears as a yellow signal, and ToTo-3 (blue) was employed for nuclear staining. CLSM images were projection images in three dimensions. Signals generated by ALP-Fast Red and HRP-DAB (**Fig. 5A–C**) are very weak compared with the fluorescent signal intensity from the QD labeling (**Fig. 5D–F**). HER2 and EGFR positive signals are distributed at the cell membranes but are barely visible in two dimensions (**Fig. 5A–C**), however cell surface signals are clearly visible in the three dimensional images generated by CLSM-META (**Fig. 5D–F**). In addition, the optical stability of the QD signals ensures adequate time for image capture.

Fig. 4. Combined observation of multiple quantum dot (QD) probes in human breast cancer tissue. Three colors were detected simultaneously: anti-rabbit-IgG-QD605 (red), anti-mouse-IgG-QD565 (green), and ToTo-3 (blue; nuclear staining). Conjugates were mixed together for multicolor fluorescence detection. Using the LSM510META system, all three colors were visible (**[A]** HER2; **[B]** ER; **[C]** nuclei; **[D]** merged image). When the same cells were examined with individual spectrum pattern analysis (**E**), only the cells stained using the QD conjugates with emissions in the META detection spectrum were visible. Note that the color differences between QD565, QD605 and ToTo-3

(*Continued on opposite page*)

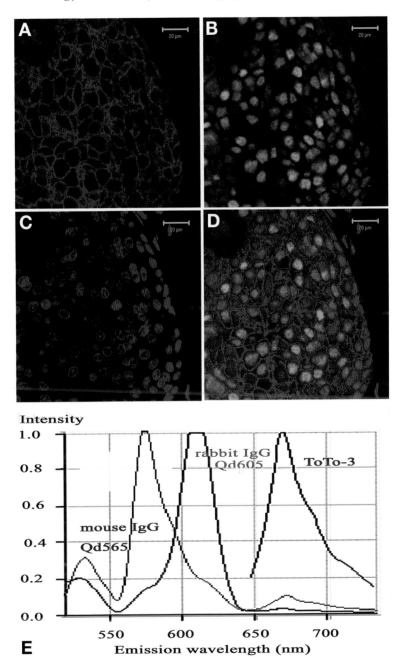

(emission peak 675 nm) were not obvious by conventional fluorescence microscopic imaging (data not shown), but were well distinguished from each other visually and spectrally by confocal laser scanning microscopy-META analysis. Bars = 20 μm.

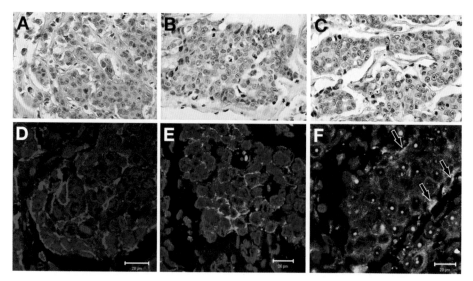

Fig. 5. Comparison of immunohistochemical detection in human breast tissue using enzyme-labeled antibodies and light microscopy (**A–C**), vs quantum dot (QD)-labeled antibodies and confocal laser scanning microscopy with META analysis (**D–E**). (**A,D**) Human epidermal growth factor receptor (HER) 2 expression. (**B,E**) Epidermal growth factor receptor (EGFR) expression. (**C,F**) Merged images of **A** and **B**, and **D** and **E**, respectively. This figure demonstrates the intensity of QD signals compared with immunoenzyme-labeled antibodies. Arrows indicate colocalization of HER2 and EGFR. Bars = 20 μm.

3.4. Conclusion

We have demonstrated that QD probes have several advantages over conventional fluorophores, and in addition have shown the usefulness of combining QD probes with CLSM-META. We have generated single fluorescent emission wavelengths from QD probes using multiple excitation lasers, and detected multiple QD signals using a signal excitation source. In addition, weak signals produced by conventional immunofluorescence and/or enzyme-labeled antibody methods have been significantly enhanced using QD labeling, and comparable signals from QDs in the same specimen area have been detected in both transmittance and META modes. The wide application of the fluorescent antibody method is now expanded to include clear detection of conventionally weak signals. We have also succeeded in electron microscopic observation of an immunofluorescence antibody reaction product that is very difficult with standard techniques. In summary, we believe that QDs represent a breakthrough for the fluorescent antibody method, expanding its already very wide application, and have demonstrated the potential of QDs for the observation of fluorescence microscopy, CLSM, and immunoelectronmicroscopy.

References

1. Itoh, J., Osamura, R. Y., and Watanabe, K. (1992) Subcellular visualization of light microscopic specimens by laser scanning microscopy and computer analysis: a new application of image analysis. *J. Histochem. Cytochem.* **40,** 955–967.
2. Espeel, M. and Van Limbergen, G. (1995) Immunocytochemical localization of peroxisomal proteins in human liver and kidney. *J. Inherit. Metab. Dis.* **18,** 135–154.
3. Kurabuchi, S. and Tanaka, S. (1997) Immunocytochemical localization of prohormone convertases PC1 and PC2 in the anuran pituitary gland: subcellular localization in corticotrope and melanotrope cells. *Cell Tissue Res.* **288,** 485–496.
4. Lontay, B., Serfozo, Z., Gergely, P., Ito, M., Hartshorne, D. J., and Erdodi, F. (2004) Localization of myosin phosphatase target subunit 1 in rat brain and in primary cultures of neuronal cells. *J. Comp. Neurol.* **478,** 72–87.
5. Ren, Y., Kruhlak, M. J., and Bazett–Jones, D. P. (2003) Same serial section correlative light and energy-filtered transmission electron microscopy. *J. Histochem. Cytochem.* **51,** 605–612.
6. Nisman, R., Dellaire, G., Ren, Y., Li, R., and Bazett–Jones, D. P. (2004) Application of quantum dots as probes for correlative fluorescence, conventional, and energy-filtered transmission electron microscopy. *J. Histochem. Cystochem.* **52,** 13–18.
7. Boisvert, F. M., Hendzel, M. J., and Bazett–Jones, D. P. (2000) Promyelocytic leukemia (PML) nuclear bodies are protein structures that do not accumulate RNA. *J. Cell Biol.* **148,** 283–292.
8. Boisvert, F. M., Kruhlak, M. J., Box, A. K., Hendzel, M. J., and Bazett-Jones, D. P. (2001) The transcription coactivator CBP is a dynamic component of the promyelocytic leukemia nuclear body. *J. Cell Biol.* **152,** 1099–1106.
9. Bruchez, M. P., Moronne, M., Gin, P., Weiss, S., and Alivisatos, A. P. (1998) Semiconductor nanocrystals as fluorescent biological labels. *Science* **281,** 2013–2016.
10. Chan, W. C. W. and Nie, S. (1998) Quantum dot bioconjugates for ultrasensitive nonisotopic detection. *Science* **281,** 2016–2018.
11. Goldman, E. R., Anderson, G. P., Tran, P. T., Mattoussi, H., Charles, P. T., and Mauro, J. M. (2002) Conjugation of luminescent quantum dots with antibodies using an engineered adaptor protein to provide new reagents for fluoroimmunoassays. *Anal. Chem.* **74,** 841–847.
12. Lingerfelt, B. M., Mattoussi, H., Goldman, E. R., Mauro, J. M., and Anderson, G. P. (2003) Preparation of quantum dot-biotin conjugates and their use in immunochromatography assays. *Anal. Chem.* **75,** 4043–4049.
13. Hanaki, K., Momo, A., Oku, T., et al. (2003) Semiconductor quantum dot/albumin complex is a long-life and highly photostable endosome marker. *Biochem. Biophys. Res. Commun.* **302,** 496–501.

14. Goldman, E. R., Clapp, A. R., Anderson, G. P., et al. (2004) Multiplexed toxin analysis using four colors of quantum dot fluororeagents. *Anal. Chem.* **76,** 684–688.
15. Gu, H., Zheng, R., Zhang, X., and Xu, B. (2004) Facile one-pot synthesis of bifunctional heterodimers of nanoparticles: a conjugate of quantum dot and magnetic nanoparticles. *J. Am. Chem. Soc.* **126,** 5664–5665.
16. Itoh, J. and Osamura, R. Y. (2004) Applications of Q dots for Bio-imaging, at light, 3D and electron microscopic levels. *J. Histochem. Cystochem.* **52,** S18.
17. Wu, X., Liu, H., Liu, J., et al. (2003) Immunofluorescent labeling of cancer marker Her2 and other cellular targets with semiconductor quantum dots. *Nature Biotechnol.* **21,** 41–46.
18. Watson, A., Wu, X., and Bruchez, M. (2003) Lighting up cells with Quantum Dots. *Biotechniques* **34,** 296–303.
19. Mansson, A., Sundberg, M., Balaz, M., et al. (2004) In vitro sliding of actin filaments labelled with single quantum dots. *Biochem. Biophys. Res. Commun.* **314,** 529–534.
20. Goldman, E. R., O' Shaughnessy, T. J., Soto, C. M., et al. (2004) Detection of proteins cross-linked within galactoside polyacrylate-based hydrogels by means of a quantum dot fluororeagent. *Anal. Bioanal. Chem.* **380,** 880–886.
21. Gao, X. and Nie, S. (2004) Quantum dot-encoded mesoporous beads with high brightness and uniformity: rapid readout using flow cytometry. *Anal. Chem.* **76,** 2406–2410.
22. Itoh, J., Yasumura, K., Takeshita, T., et al. (2000) Three-dimensional imaging of tumor angiogenesis. *Anal. Quant. Cytol. Histol.* **22,** 85–90.
23. Itoh, J., Umemura, S., Hasegawa, H., et al. (2003) Simultaneous detection of DAB and methyl green signals on apoptotic nuclei by confocal laser scanning microscopy. *Acta Histochemica et Cytochemica* **36,** 367–376.
24. Yasuda, M., Itoh, J., Kotajima, S., et al. (2004) Cytologic three-dimensional imaging for the interpretation of staining profiles: application of confocal laser scanning microscopy. *Diagnostic Cytopathology* **31,** 166–168.
25. Dickinson, M. E., Bearman, G., Tille, S., Lansford, R., and Fraser, S. E. (2001) Multi-spectral imaging and linear unmixing add a whole new dimension to laser scanning fluorescence microscopy. *BioTechniques* **31,** 1272–1278.
26. Weisshart, K., Jüngel, V., and Briddon, S. J. (2004) The LSM 510 META - ConfoCor 2 system: an integrated imaging and spectroscopic platform for single-molecule detection. *Curr. Pharma. Biotechnol.* **5,** 135–154.
27. Nakane, P. K. (1975) Recent progress in the peroxidase-labeled antibody method. *Ann. NY Acad. Sci.* **30,** 203–211.

4

Light and Electron Microscopic Localization of Multiple Proteins Using Quantum Dots

Thomas J. Deerinck, Ben N. G. Giepmans, Benjamin L. Smarr, Maryann E. Martone, and Mark H. Ellisman

Summary

Our understanding of basic cell structure and function has been greatly aided by the identification of proteins at the ultrastructural level. However, the current methods for high-resolution labeling of proteins *in situ*, and for directly correlating observations made by light microscopy (LM) and electron microscopy (EM) although invaluable, have a number of substantial limitations. These range from poor label penetration, difficulty to perform simultaneous multiprotein labeling, or the need to take the samples all the way to the electron microscope to evaluate labeling efficacy. Here we demonstrate an approach using quantum dots for pre-embedding immunolabeling of multiple diverse proteins for both LM and EM that overcomes many of these problems.

Key Words: Quantum dot; light microscopy; fluorescence microscopy; electron microscopy; immunolabeling; double labeling.

1. Introduction

Quantum dots (QDs) are fluorescent nanometer-scale semiconductor crystals that consist of an electron-dense core composed of cadmium selenide or cadmium telluride with an encapsulating shell of zinc sulfide and a polymer coat in order to facilitate bioconjugation. QDs are becoming increasingly popular as cellular probes for light microscopic imaging because of their unique optical and physical properties. These include high fluorescent quantum yield, resistance to photobleaching, a large absorption cross-section, and the ability to precisely tune the fluorescence emission by varying size, shape, and composition *(1–3)*. In addition, QDs have sufficient electron density and size diversity to be readily distinguished by electron microscopy (EM). Recently, a wide variety of

From: *Methods in Molecular Biology, vol. 374: Quantum Dots: Applications in Biology*
Edited by: M. P. Bruchez and C. Z. Hotz © Humana Press Inc., Totowa, NJ

QDs conjugated to secondary antibodies suitable for multiple labeling have become available *(4–7)*. While most of the attention has been focused on QD applications for live-cell imaging, their utility for protein labeling in fixed cells and tissues has yet to be fully characterized and exploited.

Traditionally, protein localization at the EM level has relied mainly on labeling with antigen-specific antibodies. These are then detected with secondary antibodies conjugated to labels directly by EM (up to three sizes of colloidal gold *[8]*) or to probes that can generate an electron-dense precipitate: peroxidase *(9)*, nanogold *(10)* or some fluorescent labels *(11)*. Alternatively to immunolabeling, some markers are genetically targeted *(12)*. Both colloidal gold and peroxides reaction labels have limitations such as poor penetration, complicated specimen preparation (especially for multiple labeling), or reaction product diffusion.

Antibody conjugates of QDs exhibit modest but significant penetration into fixed and mildly permeablized cells and tissues, making them suitable for pre-embedding labeling methods. The unique ability to distinguish different QDs by both light microscopy (LM) and EM makes them excellent candidates for multiple labeling as well as correlated imaging experiments *(6)*. This feature allows for efficient troubleshooting and optimization of various antibody–antigen-dependent labeling parameters by LM prior to proceeding with the more time consuming and laborious preparation for EM.

2. Materials

2.1. Labeling of Cultured Cells

1. Rat fetal lung fibroblasts (RFL-6; American Type Culture Collection, Manassas, VA) in poly-D-lysine-coated glass-bottom dishes (MatTek Corp., Ashland, MA, cat. no. P35GC-0-14-C). The cells are grown in Ham's F12 (Invitrogen, Carlsbad, CA) supplemented with 20% fetal bovine serum, 0.1 U/mL penicillin, and 0.1 mg/mL streptomycin at 37°C in 5% CO_2.
2. 1X Phosphate-buffered saline (PBS), pH 7.2, ice-cold (*see* **Subheading 2.2.**, **item 3**).
3. PEM buffer (37°C): 0.1 M PIPES, 2 mM EGTA, and 1 mM $MgSO_4$, pH 6.9.
4. Fixative (37°C): 4% formaldehyde/0.1% glutaraldehyde in PBS (*see* **Subheading 2.2.**, **item 5**; *see* **Note 1**).
5. PBS with 20 mM glycine, ice cold.
6. Ice-cold permeabilizating/blocking no. 1: 1% bovine serum albumin (BSA) (Sigma, St. Louis, MO), 0.1% Triton X-100 (Sigma) in PBS (*see* **Note 2**).
7. Ice-cold permeabilizating/blocking no. 2: 1% BSA, 0.5% Triton X-100 in PBS (*see* **Note 2**).
8. 0.1 M Sodium cacodylate buffer (*see* **Subheading 2.3.**, **item 7**).

2.2. Fixation and Tissue Preparation

1. Vibratome (Leica Corp., Wetzlar, Germany).
2. Rat Ringer's solution (pH 7.2): add the following components (all from Sigma) from previously made stock solutions: 2.5 mL Na_2HPO_4 (from 18 g/L stock),

24.8 mL NaCl (from 79.8 g/L stock), 2.5 mL KCl (from 37.5 g/L stock), 2.5 mL $MgCl_2 \cdot 6H_2O$ (from 20.0 g/L stock), 6.25 mL $NaHCO_3$ (from 50 g/L stock. Add 0.5 g dextrose. Bring total volume to 250 mL. Bubble 95% air/5% CO_2 through the solution at 35°C for 5 min. Add 2.5 mL $CaCl_2 \cdot 2H_2O$ (from 30 g/L stock). Optionally blood flow may be improved and clotting further prevented with the addition of 2.5 mL of 2% xylocaine and 25,000 U heparin.

3. 5X Stock PBS (to make 1 L): 51.1 g Na_2HPO_4, 16.8 g NaH_2PO_4, and 45.0 g NaCl, adjust to pH 7.2.
4. 4% Formaldehyde made from depolymerized paraformaldehyde (Electron Microscopy Sciences, Ft. Washington, PA) in PBS, made fresh as required prior to perfusion (*see* **Note 3**). To make 4% formaldehyde, heat 40 mL double-distilled water (ddH_2O) to 60°C and add 4 g paraformaldehyde. Add 2 drops 1 *N* NaOH and allow solution to clear. Filter through a no. 1 Whatman filter. Add 20 mL 5X PBS pH 7.2. Check pH and adjust to 7.2 with 1 *N* HCl. Bring final volume to 100 mL with ddH_2O. **Note:** toxic vapors, prepare in fume hood.
5. 4% Formaldehyde made as in **step 4** with 0.4 mL of 25% EM-grade glutaraldehyde (Tousimis, Rockville, MD) to yield 4% formaldehyde and 0.1% glutaraldehyde in PBS (*see* **Note 1**). **Note:** toxic vapors, prepare in fume hood.

2.3. Staining

1. PBS as in **Subheading 2.1.**, **item 2**.
2. Blocking/permeablization/incubation buffer: 3% of normal goat serum (Jackson, West Grove, PA), 0.1% of Triton X-100, 1% cold water fish gelatin (Sigma), and 20 m*M* glycine in PBS, chilled to 4°C for use.
3. Working buffer: 10X blocking buffer diluted in PBS, chilled to 4°C before use.
4. Primary antibodies.
5. QD-conjugated secondary antibodies (*see* **Subheading 2.4.**, **item 2**; 4 n*M* final concentration).
6. QD buffer: 2% BSA and 0.1% Triton X-100 in 50 m*M* borate buffer (pH 8.3) with 0.05% sodium azide (Quantum Dot Corporation, Hayward, CA).
7. 0.1 *M* Sodium cacodylate buffer (Ted Pella Inc., Redding, CA) from 0.3 *M* stock, adjust to pH 7.4 with 1 *N* HCl. **Note:** contains arsenic and is extremely toxic.
8. 10 mg/mL Hoechst 33342 nuclear stain (Molecular Probes, Eugene, OR), store at 4°C and protect from light.

2.4. Primary Antibodies and QD Conjugates

1. On cultured cells we used rabbit anti-Cx43 (C6219) and mouse anti-α-tubulin (T5168) from Sigma. For brain tissue we used mouse-anti-GFAP (G3895) from Sigma and rabbit-anti-IP$_3$R antibody T210 was the kind gift of T. Sudhof (*see* **Note 4**).
2. All QD reagents (Qdot® 525, 565, 585, 605, 655, and 705 nanocrystals) conjugated to goat F(ab′)2 anti-rabbit IgG and anti-mouse IgG conjugates were kindly provided by R.L. Ornberg (Quantum Dot Corp.).

2.5. Postfixation and Embedding

1. Osmium tetroxide (Electron Microscopy Sciences). Store at 4°C as a 4% aqueous stock solution. Highly toxic vapors, use only in fume hood.
2. 2% EM-grade glutaraldehyde (from 25% stock solution) in 0.15 M sodium cacodylate, pH 7.4.
3. 2% Uranyl acetate (Ted Pella Inc.) in ddH$_2$O. Filter before use. Store at room temperature. Toxic.
4. Microscope slides previously dipped in liquid release agent (Electron Microscopy Sciences, cat. no. 70880) and air-dried for longer than 24 h (slides can be stored longer than 12 mo).
5. Absolute ethanol and dry acetone.
6. Durcupan ACM resin (Electron Microscopy Sciences) is made fresh as needed by combining components A, B, C, and D by weight in the ratio 10:10:0.3:0.1, and stirring thoroughly until well mixed.
7. Three-fourth-inch paper binder clips (Office Depot, Delray Beach, FL, cat. no. OD825182) or equivalent.
8. Ultramicrotome (Leica).

3. Methods

Optimal labeling conditions must be determined for each primary antibody with the realization that four parameters (fixation, permeablization, temperature, and time) are manipulated to achieve a compromise between good staining and sufficient preservation of ultrastructure. Although no single universal protocol exists for detecting proteins with antigen-specific antibodies with adequate ultrastructural preservation for EM, starting point reference protocols for immunolabeling of cultured cells and for tissue sections along with specific guidelines are given. In general, the strongest fixation and the lowest concentration of detergent that still gives good labeling should be used (*see* **Notes 1** and **2**). One of the major advantages of using QDs is that the quality of the labeling can be assessed at the light microscopic level prior to preparation for EM, thus saving valuable time.

Six types of QDs were evaluated and are referred to by their peak fluorescence emission wavelengths. Under ultraviolet illumination, nanomolar solutions of QDs 525, 565, 585, 605, 655, and 705 nm exhibit strong and distinct narrow-band fluorescence (*see* **Fig. 1A**). When applied to a thin carbon film, air-dried, and imaged by EM they each possess a distinct size and shape profile ranging from 3 to 10 nm, with a transition from round-to-rod-shaped occurring between 585 and 605 (*see* **Fig. 1B**). Size measurements indicate that at least several combinations can be used for multiple-labeling EM experiments (*see* **Fig. 1C**). For double immunolabeling we employed 565- and 655-nm QDs with Hoechst 33342 as a nuclear counterstain. For triple immunostaining 525, 565 (or 585), and 655-nm QDs can be used (*6*), and other combinations are certainly possible.

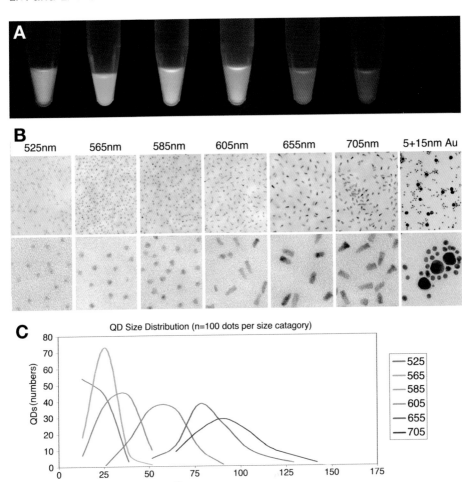

Fig. 1. Comparison of 525, 565, 585, 605, 655, and 705-nm quantum dots (QDs). (A) Fluorescence emissions of 4 n*M* solutions of QDs following ultraviolet illumination. (B) QDs plus 5- and 15-nm colloidal gold were imaged by conventional transmission electron microscopy. Identical imaging conditions and electron exposures were made at 80 keV. (C) Size distribution (length × width) of QDs from samples in B is shown (n = 100 per type).

3.1. Fixation, Permeabilization, and Staining of Cultured Cells

1. Rinse cells quickly in 37°C 1X PEM.
2. Fix at 37°C and directly transfer to 4°C for 15 min.
3. Rinse three times for 2 min at 4°C with PBS.
4. Block fixation in PBS-glycine at 4°C for 5 min.
5. Incubate in permeabilizing/blocking buffer 1 at 4°C for 15 min.

6. Rinse once in PBS at 4°C.
7. Incubated in the primary antibodies diluted with permeabilizing/blocking buffer 2 for 1–3 h at 4°C.
8. Rinse five times for 2 min at 4°C with PBS.
9. Incubated in the secondary antibodies in permeabilizing/blocking buffer 1 at 4°C overnight.
10. Rinse five times for 2 min at 4°C with PBS.
11. Rinse three times for 1 min in 0.1 M cacodylate buffer (pH 7.4).
12. If desired the cell nuclei can be counterstained with Hoechst 33342 nuclear stain diluted 1:250 from stock in 0.1 M cacodylate buffer for 5 min at 4°C (*see* **Note 5**).
13. Examine by fluorescence microscopy (*see* **Subheading 3.4.**; **Fig. 2A**). If the staining looks acceptable, cells can be further processed for EM as described in **Subheading 3.5.**, with the exception that the final dehydration is in ethanol, infiltration is in 50% Durcupan/50% ethanol (as opposed to 50% acetone), and final embedding is performed in the MatTek dish, not on specially prepared slides.

3.2. Perfusion Fixation and Vibratome Sectioning of Tissue

1. The rodent is perfused via intracardial vascular perfusion (*see* **Note 6**) with normal Ringer's solution for 2–4 min to remove blood, followed by fixative solution containing 4% formaldehyde plus 0.1% glutaraldehyde in PBS.
2. Desired tissue is removed and postfixed in 4% formaldehyde (without glutaraldehyde) in PBS at 4°C for 1 h.
3. The tissue is cut into 80- to 100-μm thick sections in 4°C PBS using a Vibratome (*see* **Note 7**).

3.3. Tissue Permiablization and Immunostaining

1. Free-floating sections are rinsed three times for 5 min in PBS and then are placed in blocking buffer for 30 min.
2. Dilutions of primary antibodies are made with working buffer (*see* **Subheading 2.3., item 3**) during the 30 min wash and kept at 4°C.
3. Sections are incubated overnight in primary antibody dilution on a shaker platform at 4°C.
4. Sections are rinsed five times for 5 min in 4°C working buffer.
5. Tissue sections are incubated in the secondary antibody-QD solution for 2–4 h on a shaker platform at 4°C, and then they are rinsed five times for 5 min with 4°C working buffer.
6. Tissue sections are rinsed three times for 10 min in 0.15 M cacodylate buffer at 4°C.

3.4. Light Microscopy

1. Some of the tissue sections can be mounted on glass slides in PBS (*see* **Note 8**) and observed at this point using a variety of fluorescence-imaging methods including conventional epifluorescence, confocal, and two-photon microscopy equipped

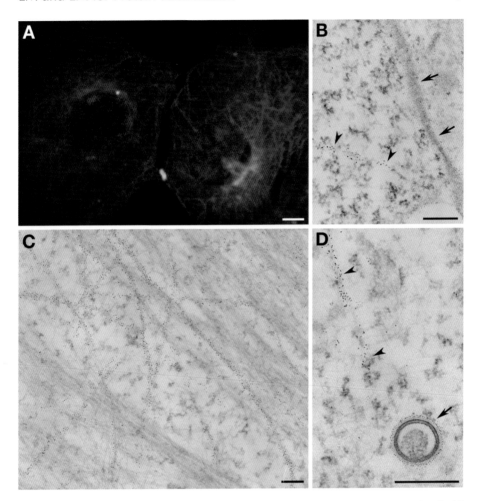

Fig. 2. Cultured fibroblasts immunolabeled for the gap junction protein Cx43 (QD 565), microtubules (QD 655), and counterstained with Hoechst 33342. **(A)** Confocal image showing labeling of microtubules (red) and a gap junction between two cells (green) with cell nuclei counterstained blue. **(B)** High magnification EM showing a QD-labeled gap junction (arrows) and a microtubule (arrowheads). **(C)** Lower magnification showing labeling of numerous microtubules stained with QD 655 near the margin of the cell. **(D)** Higher magnification showing a Cx43 annular junction and a microtubule. **(A)** Bar = 1 μm, **(B–D)** Bar = 200 nm.

with the appropriate filter sets (*see* **Note 9**). Penetration of primary and secondary antibodies can be assessed by collecting an optical section z-series (*see* **Fig. 3A,B**). If the staining and penetration look acceptable, the sections can be further processed for EM (*see* **Subheading 3.5.**).

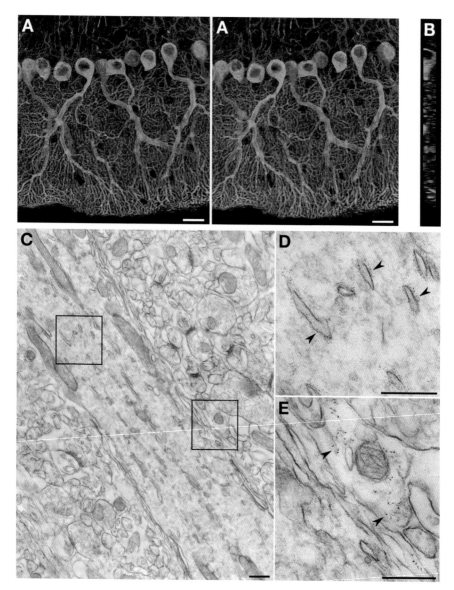

Fig. 3. Mouse cerebellum stained with anti-IP$_3$R antibodies and QD 565 (green fluorescence staining the Purkinje cells with their large branching dendrites), and anti-GFAP with QD655 (red fluorescence staining glial cells; after reference 6). **(A)** Stereopair XY image and an XZ section; **(B)** shows the penetration of the primary antibodies and QDs are greater than 5 μm. **(C)** Low-magnification image of an ultrathin section of the same preparation showing the labeling with QD 565 in the endoplasmic reticulum of a Purkinje cell dendrite (*see* enlargement in **D**) and QD 655 in a glial process **(E)**. **(A)** Bar = 10 μm, **(B)** total width of image is 10 μm, **(C–E)** Bar = 500 nm.

3.5. Processing Cells and Tissue Sections for EM

1. Postfix in 2% glutaraldehyde in 0.15 *M* cacodylate buffer on ice for 10 min.
2. Rinse three times for 5 min in 0.15 *M* cacodylate buffer and postfix in 1% osmium tetraoxide in 0.15 *M* cacodylate buffer for 30 min and then rinsed three times for 5 min in ice-cold ddH$_2$O.
3. Stain in 2% uranyl acetate in ddH$_2$O for 1 h at 4°C (*see* **Note 10**).
4. Dehydrate in a graded ethanol/ddH$_2$O series 20, 50, 70, 90, 100, and 100% on ice for 5 min each, then changed to 100% room temperature acetone. **Important:** for cells in MatTek dishes, use 100% ethanol instead. Do not use acetone, as it will dissolve the dish!
5. Infiltrate in a 50:50 dry acetone/Durcupan resin mixture (or 50:50 absolute ethanol/Durcupan for cells in MatTek dishes) at room temperature for 1–2 h, and then infiltrated twice for 1 h in 100% Durcupan resin.
6. Individual tissue sections are placed on liquid-release agent-coated embedding slides (*see* **Subheading 2.5.**, **item 4**) and a second coated slide is placed over the tissue shifted slightly lengthwise so that the slides do not completely overlap (facilitating separation after polymerization) and clamped into place using two three-quarter-inch paper binder clips (*see* **Subheading 2.5.**, **item 7**) or equivalent at the far corners of the slides, sandwiching the tissue in a thin layer of Durcupan (*see* **Note 11**). Slides or MatTek dishes are then placed in an oven at 60°C for 48–72 h, after which it can be kept at room temperature indefinitely prior to ultrathin sectioning.

3.6. Ultrathin Section Preparation and EM

1. Following polymerization, the sandwiched glass slides are carefully separated using a razorblade and the region of interest is cut out and mounted on a dummy block (Ted Pella, Inc., cat. no. 111) using crazy glue. For cultured cells the bottom cover slip is carefully removed (*see* **Note 12**) and the region of interest is removed using a jeweler's saw and glued to a dummy block.
2. After trimming, standard ultrathin sections of the vibratome section surface can be cut using an ultramicrotome and mounted on copper grids.
3. Specimens can be viewed using conventional transmission electron microscopy (*see* **Fig. 2B,C** and **Fig. 3B,C**).

4. Notes

1. The optimal fixation and permeablization conditions must be determined for each primary antibody. As a general guideline, formaldehyde can be varied from 4% down to 1%, and glutaraldehyde from 1% down to 0.05% (or omitted entirely if required). Many antigens cannot tolerate glutaraldehyde fixation, but the addition of even small amounts can improve ultrastructural preservation.
2. The detergent permeablization steps will have a significant effect on the penetration of primary and secondary antibodies. Triton X-100 can be varied from 0.01% up to 1%, with the higher concentrations having an increasingly deleterious effect on the ultrastructure. In some instances a gentler permeablization using saponin

may be substituted. It is important to note that the protocols given here are not optimal if only LM is to be performed.

3. Formaldehyde (from depolymerized paraformaldehyde) should be used within 24 h. Formalin solutions cannot be substituted.

4. The optimal concentrations for each antibody should be determined by serial dilution. For multiple-labeling experiments, each antibody should be from differing species (i.e., mouse, rabbit, and so on). For triple immunolabeling a third secondary antibody (biotinylated) and streptavidin QDs can be used (*6*).

5. Hoechst 33342 precipitates in PBS.

6. The rodent is perfused with Ringer's solution until the atrial outflow is clear and the liver has completely blanched (2–4 min for rat, 1–2 min for mouse). The fixative solution is then perfused into the animal for 3–4 min. Care should be taken to note the exact perfusion time, as it will have an significant effect on preservation and reagent penetration. With the exception of the brain, most tissues can also be adequately removed via biopsy without whole-body perfusion.

7. During Vibratome sectioning (as well as throughout most of the protocol) it is important to keep the tissue and tissue sections cold. As sections are cut they are transferred to the same fixation solution as was used during the initial fixation (1–4% formaldehyde without glutaraldehyde).

8. Tissue sections are carefully placed flat on a clean glass slide, the excess buffer is removed, and a cover slip is added (avoid air bubbles and aspirate excess fluid from around edges of the cover slip). Inspecting the tissue using LM can save valuable time from EM preparation by identifying which conditions are adequately labeled and only processing those specimens. After imaging the slide is immersed in PBS and the cover slip can be gently removed and the sections demounted. Sections not intended for EM processing can be mounted using an antifade media (i.e., gelvatol mounting medium) for long-term storage at 4°C.

9. Because of their large similar absorption cross-sections, excitation of QDs can be performed at low wavelength (e.g., 430 or 488 laser) or using two-photon excitation. Imaging shown here was performed using either a Bio-Rad Radiance confocal system or a BioRad RTS-2000MP multiphoton system.

10. At this step the sections or cells can be left in a refrigerator overnight in uranyl actetate without negatively affecting the samples.

11. This step is particularly important to ensure flatly embedded sections because the best ultrathin sections will come from the first few microns of the embedded section surface. **Caution:** clamping with excessive force can irreparably damage the section.

12. The glass cover slip can be carefully removed using a razorblade and a dissecting stereoscope. Care should be taken not to damage the epoxy surface. This process may require practice.

Acknowledgments

The authors would like to thank Ying Jones for her excellent technical assistance. This work was supported by National Institutes of Health grants NS 14718 and NCRR RR04050 to MHE.

References

1. Chan, W. C., Maxwell, D. J., Gao, X., Bailey, R. E., Han, M., and Nie, S. (2002) Luminescent Quantum dots for multiplexed biological detection and imaging. *Curr. Opin. Biotechnol.* **13**(1), 40–46.
2. Michalet, X., Pinaud, F. F., Bentolila, L. A., et al. (2005) Quantum dots for live cells, in vivo imaging, and diagnostics. *Science* **307**(5709), 538–544.
3. Bruchez, M., Jr., Moronne, M., Gin, P., Weiss, S., and Alivisatos, A. P. (1998) Semiconductor nanocrystals as fluorescent biological labels. *Science* **281**, 2013–2016.
4. Chan, W. C. and Nie, S. (1998) Quantum dot bioconjugates for ultrasensitive nonisotopic detection. *Science* **281**, 2016–2018.
5. Nisman, R., Dellaire, G., Ren, Y., Li, R., and Bazett-Jones, D. P. (2004) Application of Quantum dots as probes for correlative fluorescence, conventional, and energy-filtered transmission electron microscopy. *J. Histochem. Cytochem.* **52**(1), 13–18.
6. Giepmans, B. N., Deerinck, T. J., Smarr, B. L., Jones, Y. Z., and Ellisman, M. H. (2005) Correlated light and electron microscopic imaging of multiple endogenous proteins using Quantum dots. *Nat. Meth.* **2**(10), 743–749.
7. Giepmans, B. N., Adams, S. R., Ellisman, M. H., and Tsien, R. Y., (2006) The fluorescent toolbox for assessing protien location and function. *Science* **312**, 217–224.
8. Yeh, H. I., Rothery, S., Dupont, E., Coppen, S. R., and Severs, N. J. (1998) Individual gap junction plaques contain multiple connexins in arterial endothelium. *Circ. Res.* **83**, 1248–1263.
9. Anderson, K. D., Karle, E. J., and Reiner, A. (1994) A pre-embedding triple-label electron microscopic immunohistochemical method as applied to the study of multiple inputs to defined tegental neurons. *J. Histochem. Cytochem.* **42**(1), 49–56.
10. Takizawa, T. and Robinson, J. M. (2000) FluoroNanogold is a bifunctional immunoprobe for correlative fluorescence and electron microscopy. *J. Histochem. Cytochem.* **48**(4), 481–486.
11. Decrinck, T. J., Martone, M. E., Lev-Ram, V., et al. (1994) Fluorescence photo-oxidation with eosin: a method for high-resolution immunolocalization and in situ hybridization detection for light and electron microscopy. *J. Cell Biol.* **126**, 901–910.
12. Gaietta, G., Deerinck, T. J., Adams, S. R., et al. (2002) Multicolor and electron microscopic imaging of connexin trafficking. *Science* **296,** 503–507.

5

Human Metaphase Chromosome FISH Using Quantum Dot Conjugates

Joan H. M. Knoll

Summary

Fluorescence *in situ* hybridization (FISH) is a powerful, molecular technique with a wide range of applications in medicine and biology. In medicine, FISH uses genomic and cDNA probes to determine the chromosomal position of genes and DNA sequences, which enables detection of ploidy levels and identification of subtle chromosomal rearrangements. Because of its exquisite sensitivity, FISH often enhances conventional cytogenetic analysis and it can provide either diagnostic or prognostic results for particular chromosomal disorders. To achieve the robust probe signals needed for routine clinical application, typical FISH probes consist of recombinant genomic DNA clones often covering multiple genes. These probes can be readily visualized on metaphase chromosomes as they span tens to hundreds of thousands of nucleotides. Visualization of shorter targets has been performed in many research laboratories, and will have significant advantages once they are available clinically. Toward that end, our goal is to make the signals from these shorter probes more intense. We are utilizing quantum dot conjugates to visualize single copy sequence DNA probes as short as 1500 nucleotides in length.

Key Words: FISH; chromosome; human genome sequence; single copy probe.

1. Introduction

Fluorescence *in situ* hybridization (FISH) is utilized to detect nucleic acid sequences (generally DNA) on metaphase chromosomes and interphase nuclei *(1–5)* (for reviews, *see* **ref. 6**). A fluorescent signal is detected at the hybridization site of the specific DNA probe with its homologous chromosomal target. Hybridization is detected indirectly with fluorochrome-conjugated antibodies if the probe is labeled with modified nucleotides, or directly if the fluorochrome is conjugated directly to the nucleotide.

From: *Methods in Molecular Biology, vol. 374: Quantum Dots: Applications in Biology*
Edited by: M. P. Bruchez and C. Z. Hotz © Humana Press Inc., Totowa, NJ

Small probes have a shorter chromosomal target size than conventional recombinant probes (e.g., bacterial artificial chromosomes [BACs], cosmids), resulting in fewer fluorescent labels being detected using conventional detection methods. We have developed single copy (sc) probes and utilized them for FISH. The sc probes are on the order of 100 times shorter than conventional recombinant probes *(7,8)*. This makes probe hybridization more difficult to visualize at the microscope and can be challenging to interpret especially in the presence of non-specific binding of probes to nonhomologous chromosomal substrates, which is unavoidable with certain clinical specimens. The increased brightness and photo-stability properties of quantum dots (QDs) make them especially suitable for small probes and enhance the signals from chromosome-specific hybridization.

FISH typically involves synthesizing and purifying a DNA probe or isolating a recombinant DNA clone, labeling this molecule with a fluorescent or modified nucleotide, and hybridizing the probe to fixed cells that have been deposited and denatured on a microscope slide. Currently, QD labeling is performed by indirect detection *(9)*, i.e., modified nucleic acids can be detected with QD particles coated with antibodies that have been raised against these modifications. The following protocol has been optimized for use with sc probes, however, with minor modifications, conventional recombinant probes may also be detected with this method.

2. Materials

2.1. Preparation of Human Metaphase Chromosomes From Lymphocytes or Lymphoblast Cell Line

1. Peripheral blood sample collected in sodium heparin-coated tube or lymphoblastoid cell line.
2. CHANG MF (mitogen-free) medium (Irvine Scientific, Santa Ana, CA). This medium does not require supplements.
3. Carnoy's fixative: three parts methanol to one part acetic acid. It should be prepared fresh.
4. Colcemid: a microtubule inhibitor purchased at a ready-to-use concentration of 10 µg/mL (Irvine Scientific).
5. Ethidium bromide: dissolved in sterile water at 1 mg/mL, and stored in aluminum foil wrapped tubes at 4°C.
6. Hypotonic solution: 0.075 M KCl. Prewarm solution at 37°C before using.
7. Phytohemagglutinin (PHA; Irvine Scientific) is a T-cell mitogen. One vial (45 mg) is reconstituted in 5 mL sterile water. It can be stored at 4°C for up to 1 mo.
8. Thymidine block solution: stock solution is 10^{-1} M dissolved in sterile water and stored at −15°C.
9. Thymidine release medium: stock solution is 10^{-3} M dissolved in sterile water and stored in 1-mL aliquots at −15°C. Prior to use, add 1mL of stock to 100 mL of Chang MF medium and use 10 mL of this mixed medium for each culture.
10. Precleaned microscope slides.

11. Sterile 15-mL centrifuge tubes, glass Pasteur pipets, and rubber bulbs.
12. Equipment: tabletop centrifuge, steaming water bath, phase-contrast microscope, and a 37°C CO_2 incubator.

2.2. Probe Labeling by Nick Translation and Ethanol Precipitation

1. Purified double-stranded DNA probe prepared by long PCR amplification or cloned in a vector.
2. DNase I (Worthington Biochemical Corporation, Lakewood, NJ): stock solution is 1 mg DNase I dissolved in 1 mL 0.15 M NaCl/90% (v/v) glycerol. Store at –20°C in 20-µL aliquots. Working solution is prepared immediately prior to use by diluting 1/250 (for 4 µg/mL working solution) to 1/1000 (for 1 µg/mL working solution) in cold sterile water. Refreeze stock solution. Discard unused DNase I working solution (see **Note 1**).
3. 5 U/mL DNA polymerase I, endonuclease-free (Roche Diagnostics).
4. 10X Nick translation buffer: 2 µL each of 100 mM dATP, dCTP, dGTP (0.2 mM final) (Roche Diagnostics), 1 µL of 100 mM dTTP (0.1 mM final) (Roche Diagnostics), 500 µL 1 M Tris-HCl, pH 7.8 (0.5 M final), 50 µL 1 M MgCl$_2$ (0.05 M final), 7 µL 2-β mercaptoethanol (0.1 M final), 2 µL 50 µg/mL BSA (0.1 µg/mL final), and 435 µL sterile water. Store 50-µL aliquots at –20°C.
5. Digoxigenin 11-dUTP (Roche Diagnostics, Alameda, CA).
6. 0.8% Agarose in 1X TBE: 0.089 M Tris, 0.089 M boric acid, and 0.002 M EDTA. For 1 L of 10X TBE, use 108 g Tris-base, 55 g boric acid, and 40 mL of 0.5 M EDTA (pH 8.0).
7. Sybr gold nucleic acid stain (Molecular Probes of Invitrogen Corporation, Carlsbad, CA).
8. 0.5 M EDTA, pH 8.0: add 186.1 g of disodium EDTA·2H$_2$O to 800 mL of water. Stir vigorously on a magnetic stirrer. Adjust pH to 8.0 with NaOH pellets (~20 g). Adjust volume to 1 L and autoclave.
9. Absolute ethanol.
10. Carrier DNA: sonicate salmon sperm DNA (10 mg/mL) to molecular size of 200 to 400 bp. Store in 50-µL aliquots at –20°C.
11. 3 M Sodium acetate, pH 5.2: dissolve sodium acetate·3H$_2$O in 800 mL of water. Adjust pH to 5.2 with glacial acetic acid. Adjust volume to 1 L and autoclave.
12. 10% (w/v) Sodium dodecyl sulfate in sterile water.
13. Disposable microcentrifuge tubes and pipet tips.
14. Equipment: water bath (range 15 to 70°C), micropipettors, and a microcentrifuge.

2.3. FISH

1. Chromosome denaturation solution: 70% formamide in 2X SSC. Prepare fresh for each batch of slides. Preheat to 70°C before use.
2. 20X Sodium chloride sodium citrate solution (SSC), pH 7.0: for 1 L, dissolve 175.3 g NaCl and 88.2 g sodium citrate in 800 mL of water. Adjust pH to 7.0 with NaOH and adjust volume to 1 L. Autoclave.
3. Ethanol series: 70, 80, 95, and 100%.

4. Deionized formamide (American Bioanalytical Inc., Natick, MA).
5. 2 μL Labeled probe (~100–180 ng) in 10 μL deionized formamide per 22 mm^2 hybridization.
6. Master hybridization mix: 1 mL 20X SSC (4X final), 0.5 mL 20 mg/mL nuclease-free bovine serum albumin (Roche Diagnostics; 2 mg/mL final), 1.5 mL sterile H$_2$O, and 2 mL autoclaved 50% (w/v) dextran sulfate (Amersham Pharmacia Biotech, Piscataway, NJ; MW 500,000; 20% final). Store at 4°C for up to 6 wk.
7. Posthybridization wash solutions: 200 mL 1X SSC and 200 mL 0.5X SSC.
8. Probe detection wash solutions: 100 mL 1X SSC and 50 mL 1X SSC with 0.1% Triton.
9. Qdot® 655 sheep anti-digoxigenin conjugate (Quantum Dot Corporation of Invitrogen Corporation).
10. Disposable microcentrifuge tubes, pipet tips, 22-mm^2 plastic cover slips, and Parafilm.
11. Equipment: 37°C incubator, water bath with shaker, platform shaker, Coplin jars, and a microcentrifuge.

2.4. Chromosome Counterstaining and Mounting in Antifade Medium

1. DAPI-staining stock solution: dissolve 1 mg DAPI (0.3 mM final) in 10 mL water. Add a few drops of methanol before adding water to help dissolve DAPI. Aliquot into aluminum foil-wrapped tubes and store up to a year at –20°C. Working solution: dilute stock solution 1/500 to 1/1000 in McIlvaine buffer (pH 5.5).
2. McIlvaine buffer, pH 5.5: 43 mL 0.1 M citric acid (0.06 M final), 27 mL 0.2 M dibasic sodium phosphate (Na$_2$HPO$_4$; 0.08 M final) and autoclave before storage at 4°C.
3. Phenylenediamine dihydrochloride antifade: dissolve 50 mg p-phenylenediamine dihydrochloride in 5 mL 1X phosphate-buffered saline (PBS) (9 mM final). Adjust to pH 8.0 with 0.5 M carbonate/bicarbonate buffer, pH 9.0. Add to 45 mL glycerol (90% final), mix, and filter through a 0.22-μm filter. Store in air-tight microcentrifuge tubes in 50-μL aliquots in the dark at –20°C. Thaw and use at once. Restrict exposure to light and air as the antifade will discolor.
4. PBS: prepare 10X stock with 1.37 M NaCl, 27 mM KCl, 100 mM Na$_2$HPO$_4$, 17.6 mM KH$_2$PO$_4$ (adjust pH to 7.4 with HCl if necessary) and autoclave before storage at room temperature. Prepare a 1X working solution.
5. 0.5 M Carbonate/bicarbonate buffer, pH 9.0: 0.42 g NaHCO$_3$ in 10 mL H$_2$O. Adjust pH to 9.0 with NaOH.
6. Disposable microcentrifuge tubes, micropipet tips, Parafilm, glass cover slips (no. 0 thickness), and nail polish.
7. Equipment: microcentrifuge, Coplin jars, and a 37°C incubator.

2.5. Microscopy

1. Fluorescence microscope with epi-illumination and filter set(s) appropriate for DAPI and QDot 655 fluorochrome(s) or other fluorochromes utilized (Chroma Technology Corporation, Rockingham, VT).
2. Camera and imaging system attached to microscope.

3. Methods

The metaphase FISH protocol described in this section is for detecting digoxigenin-labeled short sequence-defined DNA probes (i.e., sc- or single-copy probes), designed directly from the human genome sequence *(8)*; for additional details, (*see* http://www.scprobe.info). Although the following protocol uses digoxigenin-11-dUTP as the modified nucleotide, DNA probes can also be labeled with biotin-16-dUTP and detected with streptavidin-conjugated QDs.

3.1. Cell Culture and Preparation of Metaphase Chromosomes

1. Culture initiation. Mix 9.0 mL Chang MF media with 0.1 mL PHA (45 mg/mL) and 1 mL of peripheral blood in each 15-mL tube. Mix, cap, and incubate cultures at a 45° angle at 37°C with 5% CO_2 for 2–3 d (*see* variations below for difference in chromosomal length) (*see* **Note 2**).
2. Harvest of routine cultures (moderate chromosome length). After 72 h, add 0.1 mL ethidium bromide to each 10-mL culture for 90 min at 37°C, followed by the addition of 0.1 mL colcemid for 20 min at 37°C. Centrifuge at 300*g* for 10 min. Aspirate supernatant leaving 0.5 mL of supernatant above the pellet. Resuspend the pellet by tapping the bottom of the tube. Add 5 mL of prewarmed KCl hypotonic solution and gently pipet to mix. Add an additional 5 mL of KCl hypotonic and pipet to mix. Reincubate at 37°C for 25 min. Add 2 mL of freshly made Carnoy's fixative to one tube at a time. Recap the tube and gently invert to mix. Centrifuge at 300*g* for 10 min. Aspirate most of the supernatant from the lymphocyte pellet. Tap tube gently to mix. Slowly add up to 10 mL of fixative and gently pipet to disperse evenly. Centrifuge at 300*g* for 10 min, aspirate the supernatant, gently disperse the cell pellet, and repeat the fixative wash until cell pellet is white (usually three washes). After a final wash store cells in fixative at 4°C (*see* **Note 3**).
3. Harvest of thymidine cultures (longer chromosome length). After 48 h, add 0.1 mL of thymidine block solution, mix, and reincubate overnight at 37°C (~17.5 h). Centrifuge at 300*g* for 10 min. Remove the supernatant and resuspend the cell pellet in 10 mL of the thymidine release medium for a total of 4.5 h. Add 0.1 mL colcemid and reincubate for 20 min. Continue with hypotonic treatment and fixation as for routine cultures (*see* **Note 3**).
4. Microscope slide preparation. Remove fixed cell cultures from 4°C, centrifuge at 300*g*, and then examine the cell pellet. If the pellet appears gray or tan, wash with fixative once or twice before making the slides. Remove the supernatant from the pellet and resuspend the pellet in a small volume of fixative (sufficient to achieve a slightly cloudy suspension). Add one or two drops of cell suspension onto a cleaned slide with a thin layer of fixative. Hold the slide at an angle as the cell drops are placed on it such that the cell suspension spreads evenly over the region to be hybridized. Briefly place the slide over a steaming water bath. Inspect the slide with a phase-contrast microscope and etch the best region for hybridization (*see* **Note 4**).

3.2. Probe Labeling by Nick Translation and Ethanol Precipitation

Nick translation is a common method for labeling probes *(10,11)* with digoxigenin, biotin, and fluorochrome-tagged deoxynucleotides, but other methods such as random oligonucleotide-primed labeling are also effective *(12,13)*. Probes can be PCR products amplified from the human genome as in sc *(7,8)* or recombinant genomic DNA sequences cloned in vectors such as plasmids, phage, cosmids, P1s, P1-derived artificial chromosomes, BACs, or yeast artificial chromosomes.

1. Add the following, in the order listed to a 1.5-mL microcentrifuge tube on ice: 10 µL 10X nick translation buffer (1X final), 1 µg probe DNA, 5 µL 1 mM digoxigenin-11-dUTP (0.05 mM final), sterile H$_2$O to give a final total volume of 100 µL, 10 µL 1 µg/mL DNase I, and 4 µL 5 U/mL DNA polymerase I. Mix ingredients and microcentrifuge briefly, then incubate ≥1.5 h at 15°C to obtain the optimal incorporation of modified deoxnucleotide.

2. Remove a 10-µL aliquot of nick translation reaction and heat to 65°C for 2 min and place on ice. Store the remaining reaction volume (i.e., 90 µL) at –20°C. Check the probe size range by performing electrophoresis of a 10-µL aliquot in a 0.8% agarose minigel in 1X TBE. For efficient hybridization, the majority of extension products should be between 250 and 600 bp in length (*see* **Note 5**).

3. When correct probe size distribution is obtained, the reaction is stopped by inactivating the enzymes in the remaining reaction volume (90 µL) by the addition of 3 µL 0.5 M EDTA and 1 µL 10% SDS. The reaction may also be incubated at 65°C for 10 min.

4. Precipitate labeled probe by adding 1 µL of 10 mg/mL sonicated salmon sperm carrier DNA, 1/10 vol 3 M sodium acetate, and 2 1/2 vol cold 95% ethanol to reaction mixture. Invert and place at –20°C for at least 1 h.

5. Microcentrifuge for 30 min at maximum speed (20,000g) at 4°C. Remove supernatant, and rinse the DNA pellet with cold 70% ethanol. Air-dry and resuspend the pellet in 10 µL sterile water (~90 ng/µL). Store at –20°C until used for hybridization (*see* **Note 6**).

3.3. FISH

3.3.1. Denaturation of Chromosomal DNA and Probe DNA

Denaturation of the chromosomal and probe DNA are generally performed separately to yield chromosomes of higher quality. Co-denaturation of chromosomal and probe DNA on the microscope slide yields metaphase cells with less distinct banding and poor morphology.

1. Heat 50 mL of chromosomal denaturation solution (70% formamide in 2X SSC) in a Coplin jar to 70°C. Place the microscope slide with the cells into preheated denaturation solution for 2 min. For multiple slides, stagger denaturation at 1-min intervals (*see* **Note 7**).

2. Remove slide and rinse in cold 70% ethanol for 2 min followed by two 1-min rinses each in a room temperature ethanol series (80, 95, and 100%). Allow slide to air-dry before hybridizing with probe. Examine chromosomal morphology with phase-contrast microscopy (*see* **Note 8**).
3. For probe denaturation, mix 2 μL of labeled sc probe (~180 ng) with 10 μL deionized formamide, pool, and heat for 5 min at 70°C and then briefly place on ice (*see* **Note 9**).

3.3.2. Hybridization of Probe DNA With Chromosomal Target

1. Add 10 μL master hybridization mix (prewarmed to 37°C) to denatured probe.
2. Mix well, and spin briefly at high speed in a microcentrifuge to pool contents.
3. Pipet the entire volume of probe (~22 μL) and place over the denatured chromosome preparation on slide.
4. Cover with a 22-mm² plastic cover slip, and remove any large air bubbles with gentle pressure.
5. Seal between two sheets of Parafilm, and incubate overnight (~14–18 h) at 37°C. The Parafilm sandwich prevents evaporation.

3.3.3. Posthybridization Washes to Remove Nonspecific Probe Attachment

1. Remove slides from overnight incubation, remove parafilm seal, and carefully lift off plastic cover slips.
2. Wash hybridized slides in 1X SSC for 2 min at 50°C, followed by four rinses at 5 min each in 0.5X SSC at 50°C and three rinses at 3 min each at 50°C. For decreased wash stringency, increase the SSC concentration to 4X and decrease the temperature to 39°C. Do not allow slide to dry at any point in the procedure.

3.3.4. Detection of Hybridized Digoxigenin-Labeled Probe

1. Dilute the QD-conjugated digoxigenin antibody stock (1 μ*M*) to 5–20 n*M* using the buffer supplied by the manufacturer. Let stand for 20–30 min.
2. Apply 50 μL of the diluted antibody to the hybridized area of the slide and cover with a parafilm cover slip.
3. Seal the slides in a parafilm sandwich and incubate at 37°C for 1 h in the dark on a platform shaker set on slow speed.
4. Remove the parafilm seal and cover slips. Place slides for 20 min each in 1X SSC, 1X SSC with 0.1% Triton, followed by 1X SSC. All postdetection washes are performed at room temperature in aluminum foil-covered Coplin jars and on a platform shaker.

3.3.5. Counterstain Chromosomal DNA and Mount in Antifade

1. Add 50 μL DAPI staining solution to the hybridized chromosome preparation on the slide, and cover with 22-mm² square of Parafilm.

2. Stain for 10 min at room temperature in a light-proof container (*see* **Notes 10 and 11**).
3. Rinse briefly in 1X SSC. Blot slide, but do not dry.
4. Add 6 μL antifade mounting medium to stained slide and add a 22-mm^2 glass cover slip (no. 0 thickness).
5. Eliminate the excess antifade medium by gently squeezing the cover slip against the glass slide and seal with nail polish. Examine the slide immediately or store at –20°C in a slide box (*see* **Note 12**).

3.3.6. Microscopy and Analysis

1. Examine the slide using a fluorescence microscope with epi-illumination and filter sets appropriate for the fluorochromes used. A DAPI filter set is used for chromosome identification and a QD 655 filter set for probe hybridization. Multipass filter sets can be used for simultaneous viewing of the fluorochromes (*see* **Note 11**).
2. Photograph cells with a digital or charge-coupled device camera and computer image capture system. Intermittent emission (or blinking) of individual QD nanoparticles can influence the choice of the correct exposure time (*see* **Note 12**).
3. A typical analysis includes examination of hybridization efficiency of 20 normal metaphase cells. Efficiency is determined by scoring the frequency of hybridization on both chromosomes. This protocol has been used on probes ranging in size from 1534 bp (chromosome 9q34) to 3929 bp (chromosome 22q11.2). The detection efficiency ranged from 60 to 90% (*see* **Note 13**). The hybridizations were considerably larger than observed with rhodamine-conjugated digoxigenin-dUTP. **Figure 1** shows a representative hybridization of a sc probe.

4. Notes

1. Unless otherwise stated, all solutions should be prepared in water that has a resistivity of 18 MΩ-cm. This standard is referred to as "water" in this text. Water is sterilized by autoclaving.
2. PHA is not necessary for lymphoblastoid cell lines. Chromosomes can also be obtained from other tissue types in which cells are capable of undergoing cell division. Other tissue types have different culturing and harvesting requirements (**14**).
3. Fixed metaphase cell pellets can be stored indefinitely in screw-cap polypropylene tubes in Carnoy's fixative. The cells are resuspended in fresh fixative prior to microscope slide preparation.
4. To maximize probe hybridization efficiency, it is important to have high-quality cell preparations. For best hybridization results, unstained chromosomes should appear well spread with minimal chromosome overlap, evenly dried, nonrefractile, and dark gray or light black when examined with a phase-contrast microscope. In addition, there should be minimal extracellular debris in the preparations. Adjust drying conditions as necessary to obtain these morphological qualities.
5. Probe hybridization efficiency and background are also influenced by size of the labeled probe. Optimal probe size is 250–600 bp; background increases when this size is exceeded. If the probe extends beyond 800 bp, additional nick translation

Fig. 1. A normal human metaphase cell hybridized with a 3929-bp breakage cluster region (BCR) gene probe on chromosome 22q11.2. The probe is detected with anti-digoxigenin conjugated QDot 655. The chromosome 22s are indicated (arrows) and the hybridization signal is comparable to that of much longer cloned probes.

of the probe is recommended. If the majority of probe DNA is larger than 800 bp in size, add 5 µL of 4 µg/mL DNase I to the remaining 90 µL of reaction mix, incubate 30 min at 15°C, and check gel size again. Additional DNase I is added in a more concentrated solution to avoid a significant increase in volume. Each new lot of DNase I is tested for activitiy.

6. Labeled probes can be stored in individual use aliquots for at least 1 yr at –20°C. Because of the long shelf life, large quantities of probe DNA can be labeled by performing multiple 100-µL reactions in parallel, each containing 1 µg of probe DNA. Incorporation of labeled nucleotide can be measured utilizing commercially available colorimetric assays or radioactively. Incorporation is usually verified only when calibrating activity of new lots of DNAse I and DNA polymerase I, or if hybridization results suggest that probes are being detected inefficiently.

7. The quality of formamide, temperature, pH, and duration of exposure of the probe to denaturation solution are important parameters for successful hybridization. The denaturation solution should not be kept in the 70°C water bath longer than 30 min as the pH will drop, and chromosome morphology will be adversely affected.

8. Examine denatured chromosome preparation with a phase-contrast microscope. For the best hybridization results, the chromosomes will appear lighter than undenatured chromosomes, but will still have a recognizable internal structure along their length. Overdenaturation results in chromosomes with a hollow, transparent appearance. DAPI counterstaining of these over-denatured preparations results in poor chromosome banding and chromosomes with a "fuzzy" appearance. Incomplete denaturation results in chromosomes that appear similar to untreated preparations. Freshly made slide preparations denature quicker than those which are 2 wk or older.

9. Sequence-defined sc probes, α-satellite and telomeric probes, and most cDNA probes do not require suppression with C_0t-1 DNA. For probes that do contain repetitive elements (typically recombinant genomic DNA clones), repeat suppression is performed as follows: add 10 µg human C_0t-1 DNA to approx 100–150 ng labeled DNA probe, and vacuum-dry before resuspending in 10 µL of deionized formamide and denaturing at 70°C. Following denaturation, do not place on ice. Instead, pulse centrifuge to spin content to bottom of tube and place in a 37°C water bath for 30 min to several hours to permit preannealing of the repetitive elements with C_0t-1 DNA. Generally larger probes, such as P1, P1-derived artificial chromosomes, BACs, or yeast artificial chromosomes and those with a large amount of repetitive sequence, should be preannealed for a longer duration *(15)*.

10. The slide is ready to examine immediately after it is mounted, but it may be stored several weeks at –20°C with minimal loss of fluorescence. Frozen slides should be warmed to room temperature before viewing. Because detection reagents and DAPI are light sensitive; exposure to ambient light should be minimized by use of light-tight storage containers.

11. DAPI staining permits chromosome identification as it provides a chromosomal banding pattern comparable to the one obtained with trypsin and Geimsa stain (i.e., GTG-banding). DAPI excites at 360 nm and emits maximally at approx 460 nm; however, it has a broad emission spectrum that substantially overlaps with several of the commercially available QDs. All QDs excite with ultraviolet wavelengths, but emit at different wavelengths depending on the QD conjugate size. The Qdot 655 (which emits at 655 nm) has the advantages of both being small and exhibiting minimal bleed-through from the DAPI-stained chromosomes. Although the lower emission wavelength QDs are visible, the emission filter does not discriminate probe signals from the DAPI counterstain as well.

12. A blinking or flashing pattern from the QD nanoparticle occurs in some specimens. The frequency of this phenomenon seems to be related, in part, to cytological preparation quality. Brief exposure to the cold seems to reduce the blinking.

13. If the hybridization signal is not visible or is difficult to see, it is possible that the chromosomal target is too small, the level of nucleotide incorporation is inadequate, the concentration of probe utilized is too low, or chromosome denaturation is suboptimal. The level of incorporation of labeled nucleotide should be checked and/or the concentration of probe utilized for hybridization adjusted if necessary.

Acknowledgments

The author would like to thank Professor Peter K. Rogan for comments on this chapter, and Angela Marion for technical assistance. This work was supported by the Katherine B. Richardson Foundation, PHS R21CA095167-02, and R41CA112692-01, subcontract from Phylogenetix Laboratories, Inc. The QD reagents were a generous gift from Dr. Richard Ornberg at Quantum Dot Corporation.

References

1. Landegent, J. E., Jansen in de Wal, N., Dirks, R. W., Baao, F., and van der Ploeg, M. (1987) Use of whole cosmid cloned genomic sequences for chromosomal localization by non-radioactive in situ hybridization. *Hum. Genet.* **77**, 366–370.
2. Landegent, J. E., Jansen in de Wal, N., van Ommen, G. J., et al. (1985) Chromosomal localization of a unique gene by non-autoradiographic in situ hybridization. *Nature* **317**, 175–177.
3. Lawrence, J. B., Villnave, C. A., and Singer, R. H. (1988) Sensitive, high-resolution chromatin and chromosome mapping in situ: presence and orientation of two closely integrated copies of EBV in a lymphoma line. *Cell* **52**, 51–61.
4. Pinkel, D., Landegent, J., Collins, C., et al. (1988) Fluorescence in situ hybridization with human chromosome-specific libraries: detection of trisomy 21 and translocations of chromosome 4. *Proc. Natl. Acad. Sci. USA* **85**, 9138–9142.
5. Pinkel, D., Straume, T., and Gray, J. W. (1986) Cytogenetic analysis using quantitative, high-sensitivity, fluorescence hybridization. *Proc. Natl. Acad. Sci. USA* **83**, 2934–2938.
6. Levsky, J. M. and Singer, R. H. (2003) Fluorescent in situ hybridization: past, present and future. *J. Cell. Sci.* **116**, 2833–2838.
7. Knoll, J. H. and Rogan, P. K. (2003) Sequence-based, in situ detection of chromosomal abnormalities at high resolution. *Amer. J. Med. Genetics* **121A**, 245–257.
8. Rogan, P. K., Cazcarro, P. M., and Knoll, J. H. (2001) Sequence-based design of single-copy genomic DA probes for fluorescence in situ hybridization. *Genome Research* **11**, 1086–1094.
9. Xiao, Y. and Barker, P. E. (2004) Semiconductor nanocrystal probes for human metaphase chromosomes. *Nucleic Acids Res.* **32**, e28.
10. Rigby, P. W., Dieckmann, M., Rhodes, C., and Berg, P. (1977) Labeling deoxyribonucleic acid to high specific activity in vitro by nick translation with DNA polymerase I. *J. Mol. Biol.* **113**, 237–251.
11. Langer, P. R., Waldrop, A. A., and Ward, D. C. (1981) Enzymatic synthesis of biotin-labeled polynucleotides: novel nucleic acid affinity probes. *Proc. Natl. Acad. Sci. USA* **78**, 6633–6637.
12. Feinberg, A. P. and Vogelstein, B. A. (1984) Technique for radiolabeling DNA restriction endonuclease fragments to high specific activity. Addendum. *Anal. Biochem.* **137**, 266–267.
13. Morrison, L. E., Ramakrishna, R., Ruffalo, T. M., and Wilber, K. A. (2002) Labeling fluorescense *in situ* hybridization probes for genomic targets. In:

Molecular Cytogenetics: Protocols and Applications, (Fan, Y. S., ed.), Humana Press, Totowa, NJ, pp. 21–40.

14. Barch, M. J., Knutsen, T., and Spurbeck, J. L. (1977) *The ACT Cytogenetics Laboratory Manual. 3rd ed.*, Lippencott-Raven, Philadelphia, PA.

15. Knoll, J. H. and Lichter, P. (1994) *In situ* hybridization to metaphase chromosomes and interphase nuclei. In: *Current Protocols in Human Genetics*, (Dracopoli, N. C., Haines, J. L., Korf, B. R., et al., eds.), Green-Wiley, New York, pp. Unit 4.3.

II

IMAGING OF LIVE CELLS

6

Biotin-Ligand Complexes With Streptavidin Quantum Dots for In Vivo Cell Labeling of Membrane Receptors

Diane S. Lidke, Peter Nagy, Thomas M. Jovin, and Donna J. Arndt-Jovin

Summary

The unique fluorescence properties of quantum dots (QDs), particulary their large extinction coefficients and photostability, make them ideal probes for tracking proteins in live cells using real-time visualization. We have shown that QDs conjugated to epidermal growth factor act as functional ligands for their receptor, erbB1. Here, we describe protocols for (1) conjugation of streptavidin–QDs to biotinylated ligand, (2) formation of ligand–QD-receptor complexes, and (3) quantification of binding and internalization of receptor complex using both high-resolution fluorescence microscopy and flow cytometry.

Key Words: Epidermal growth factor; EGF; epidermal growth factor receptor; EGFR; erbB1; endocytosis; biotin; streptavidin.

1. Introduction

The erbB family of receptor tyrosine kinases includes erbB1 (the classical epidermal growth factor receptor, EGFR), erbB2, erbB3 and erbB4 *(1)*. Binding of epidermal growth factor (EGF) to the extracellular domain of erbB1 leads to auto- and transactivation of cytoplasmic protein kinase domains in response to the formation *(2)* or reorientation *(3–5)* of homo- and hetero-associated receptor tyrosine kinase monomers. This activation of the receptor initiates signaling cascades, which control a variety of cellular processes such as DNA replication and division *(1,6)*. ErbB1 activation results in clathrin-mediated endocytosis via coated pits, covalent modification of the receptor, and endosomal trafficking resulting in both downregulation through proteosomal degradation and recycling to the cell membrane. The proper function of erbB1 in the cell is critical, given that overexpression or mutations of the erbB1 are implicated in many types of cancer *(1,7)*.

From: *Methods in Molecular Biology, vol. 374: Quantum Dots: Applications in Biology*
Edited by: M. P. Bruchez and C. Z. Hotz © Humana Press Inc., Totowa, NJ

We have shown that quantum dots (QDs) conjugated to EGF act as functional ligands for erbB1, and that their unique fluorescence properties provide the means for prolonged real-time visualization of the multiple steps in signaling mechanisms in living cells *(8,9)*. In this chapter we outline procedures for binding biotinylated ligands to commercial streptavidin-conjugated QDs for specific labeling of growth factor receptors with these ligands and for quantitative measurements of the receptor complexes by both fluorescence microscopy and flow cytometry.

2. Materials

1. Phosphate-buffered saline (PBS): 137 mM NaCl, 2.7 mM KCl, 7.9 mM Na$_2$HPO$_4$, and 1.5 nM KH$_2$PO$_4$, pH 7.3. Autoclave before use and store at room temperature. PBS/bovine serum albumin (BSA): add 0.1% BSA just before use.
2. Tyrode's buffer without glucose: 135 mM NaCl, 10 mM KCl, 0.4 mM MgCl$_2$, 1 mM CaCl$_2$, and 10 mM HEPES, pH 7.2. Autoclave before use and store at room temperature. Tyrode's plus: add 20 mM glucose and 0.1% BSA just before use.
3. Streptavidin-conjugated QDs (SAvQD) from Invitrogen (Eugene, OR, www.probes.com).
4. LabTek eight- or two-well cover slip chambers from Nunc (Rochester, NY).
5. Cover slips, acid washed and sterilized.
6. Cell culture medium appropriate for the particular cell line. Various commercial vendors.
7. Emission filters for specific QD emission wavelengths, 20 nm FWHM Chroma (Brattleboro, VT).
8. Biotinylated EGF and EGF from Molecular Probes (Invitrogen, www.probes.com).
9. Paraformaldehyde (PFA) fixative: PFA, analytical grade. Freshly prepared 4% solution in PBS. A 40% solution can be prepared and kept in small aliquots at –20°C. Thaw at 50°C and immediately dilute in PBS before use.
10. Tris-saline buffer: 125 mM NaCl, 25 mM Tris-HCl, pH 7.3.
11. Trypsin solution for releasing adherent cells from their substrates: sterile solution of 0.5 mg/mL trypsin in PBS containing 1 mM EDTA.
12. Acid wash buffer: 0.5 M NaCl and 0.1 M glycine-HCl, pH 2.5.

3. Methods

3.1. Making Biotin-Ligand Linked SAvQD

3.1.1. Ratio of Biotinylated Ligand to QDs

The present formulation of the QDs from Quantum Dot Corp. includes a polyethylene glycol 2000 outer layer to which 5–10 streptavidin molecules per QD are covalently linked. This allows the stochiometry of the ligand–QD complex to be varied. It is preferable to use monovalent biotinylated ligands where the biotin is conjugated to a single known residue or position. In this way the properties of the ligand–QD complex can be better controlled and standardized, and potential cross-linking of the QDs via the ligand is avoided. By mixing SAvQDs, biotinylated ligand, and free biotin at different molar ratios, one can create QDs with different numbers of attached ligands. We assume a Poisson distribution. To create primarily monovalent biotin conjugates, mix a 10-fold molar excess of SAvQDs with the biotinylated ligand.

3.1.2. Preformed Complexes

1. Dilute SAvQD to 20 nM in PBS/BSA.
2. Dilute biotinylated ligand to the appropriate desired molarity in PBS/BSA.
3. Add SAvQDs to ligand and mix with a micropipet.
4. Incubate at 4°C for at least 30 min with gentle agitation or rotation (*see* **Note 1**).

3.1.3. Purification of Ligand–QD Conjugates

After the ligand–QD complex is formed, unbound ligands that are smaller than the QDs can be removed from the conjugate by passing the QDs through a size exclusion column. For SAvQDs described, Sephadex G-25 medium grade gel filtration is recommended; the SAvQDs will elute in the void volume. Use a 1:20 ratio of sample to column volume for good separation. Spin columns can be used for sample volumes of 50 µL or less.

1. Swell and pour the Sephadex according to the manufacturer's instructions.
2. Equilibrate the column in PBS by washing with at least three column volumes.
3. Allow the buffer to just enter the top of the gel bed before adding the sample. Avoid air bubbles.
4. Add the sample to the top of the gel bed, allow the sample to enter the gel, and add PBS to elute the sample.
5. Collect the fractions containing the QDs.

3.2. In Vivo, In Situ Labeling of Cells With Ligand–QD Conjugates for Fluorescence Microscopic Imaging

3.2.1. Plating cells—Cover Slips and Cover Slip Chambers

Logarithmically growing cells are replated in complete culture medium at the appropriate densities 1 or 2 d prior to the experiment. In the case of experiments with growth factor receptors, the cells are serum-starved for 16–24 h prior to the experiment to reduce signaling induced by the serum in the medium.

3.2.2. Labeling Cell Surface Receptors With Preformed Ligand–QD Complex

1. Wash cells once with Tyrode's plus and maintain in this buffer for the experiment.
2. Place the cover slip chamber on the microscope stage and equilibrate to the desired temperature.
3. Dilute the ligand–SAvQD complex (in this case EGF–QD) to twice the final concentration in Tyrode's plus (*see* **Note 2**).

4. Add an equal volume of EGF–QD in Tyrode's plus to the cells in the well under observation to a final concentration of 5–200 pM.
5. Start imaging.

3.2.3. Labeling Cell Surface Receptors in a Two-Step Procedure

In some situations the preformed complex may not be functional or the system may not respond properly with the preformed complex because of (1) steric hindrance by the SAvQD during ligand binding or (2) SAvQD inhibition of a conformational change of the ligand required for binding. In addition, ligands with multiple biotins can cause aggregation of SAvQDs. Thus, if one does not see binding with the preformed complex or observes aggregation of SAvQDs in the one-step protocol, the following two-step protocol may be useful. We have successfully used this procedure with both the EGF and the transferrin receptors *(8,10)*.

1. Wash cells once with ice cold Tyrode's plus and keep in this buffer for the experiment. Put cells on ice.
2. Incubate cells with biotinylated-ligand (in this case 1 nM biotin–EGF, *see* **Note 3**) for 5–10 min.
3. Wash with ice cold Tyrode's plus several times.
4. Add cold 200 pM SAvQD for 5–15 min.
5. Wash the cells several times with cold Tyrode's plus.
6. Place cells on the microscope stage with appropriate temperature control and image.
7. For snapshots of the behavior of the ligand–QD complex over time, incubate cover slips at the appropriate temperatures and times and fix in 4% PFA for 15 min on ice to prevent redistribution of the QDs. After fixation, wash the cover slips with Tris-saline buffer for 10 min several times and mount in PBS for microscopy.

3.2.4. In Vivo Visualization and Quantitation of Ligand–QD Binding by Fluorescence Confocal Microscopy

Quantitative binding and/or internalization kinetic plots can be generated from the QD signals if cells express a membrane protein fused to a visible fluorescent protein (GFP or any VFP) or are stained with a membrane-specific probe.

1. Grow cells in an eight-well LabTek chambers as previously described.
2. Wash cells once with Tyrode's plus and keep in 200–250 µL of Tyrode's plus for imaging.
3. Begin acquisition of a time series with single or multiple focal planes.
4. Typically, a 63× or 40× 1.2 NA water immersion objective is recommended. Simultaneous excitation of GFP and QD605 can be achieved at 488 nm. An imaging system with two detectors, e.g., Zeiss LSM510-META, allows the simultaneous collection of GFP and QD signals with appropriate filters (in this case, 520/20 bandpass and 585 longpass, respectively).
5. After several time-points are acquired, add the preformed EGF–QD complexes to the buffer (1:1 dilution) without disrupting the time series.

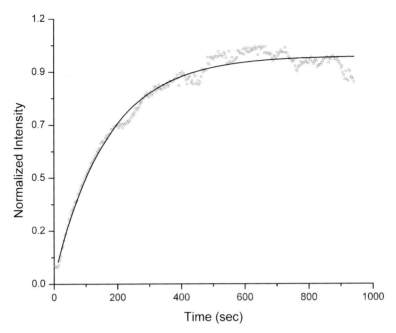

Fig. 1. Quantitation of epidermal growth factor–quantum dots (EGF–QDs) binding to Chinese hamster ovary cells expressing erbB1–eGFP. Plot shows the change in QD intensity inside the membrane mask over time. The binding of the EGF–QDs to the cell surface follows an exponential course.

6. Analysis *(8)*.
7. Background subtraction from both channels is performed.
8. The membrane region of the cell is segmented by a global threshold in the green channel and the mask is dilated according to a pixel width corresponding to the membrane circumference.
9. The average QD intensity (from the red channel) in the membrane mask is plotted over time (*see* **Fig. 1**).

3.2.5. Two-Color QD "Chase" Experiment

Multicolor EGF–QDs may be used to track endocytosis of the receptor–ligand complex and to follow subsequent steps in time. Two different colors of ligand–QDs are delivered sequentially (*see* **Fig. 2**).

1. The day before the experiment, plate cells on cover slips if taking fixed cell time-points or in Lab-Tek cover slip chambers for live cell imaging.
2. Prepare two batches of QD–ligand, with separated emission spectra, such as SAvQD525 and SAvQD605.
3. Wash the cells in Tyrode's plus buffer and keep in this buffer.

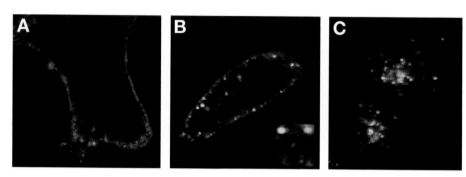

Fig. 2. Two-color quantum dot (QD) "chase" experiment of epidermal growth factor (EGF)–QDs binding to A431 cells expressing erbB1. **(A)** EGF-QD605 (red) was bound to ErbB1 on A431 cells for 3 min at room temperature. Cells were washed and incubated for 5 min 37°C to induce internalization. EGF-QD525 (green) was then added and incubated for 8 min at 37°C. Early endosomes are red only, indicating that the EGF-QD605 has internalized, whereas the EGF-QD525 remains predominantly external. **(B)** Twenty minutes later. Several different classes of endosomes are observed. The endosomes contain red only, green only or both QD colors (yellow, *see* inset), indicating that early, sorting and late endosomes are loaded. **(C)** Two hours later. Although all vesicles show red and green signals, clear differences in QD populations are seen in the content of late endosomes and lysosomes.

4. Incubate the cells in a non-saturating concentration (<200 p*M* in the case of EGF–QD) of the first QD–ligand (QD605) for 5–10 min at 37°C while taking time-lapse images on a confocal microscope or incubate cover slips in a moist chamber for samples to subsequently fix.
5. Wash away excess QD–ligand by two changes of Tyrode's plus.
6. Incubate at 37°C for at least 5 min to clear the membrane of ligand–QD (image in vivo).
7. Add 1 n*M* of ligand of another color, EGF–QD (QD525), and incubate at 37°C for desired time in vivo with imaging or for desired fixed time-points.
8. For fixed samples, wash cover slips in Tyrode's buffer and fix at 4°C for 15 min in 4% PFA (*see* **Subheading 3.2.3.**).
9. Wash with cold Tris-saline several times.
10. Antibody staining for activated erbB1, endocytic markers, or other cell components can be carried out at this point.
11. Mount cover slips onto a drop of PBS on a microscope slide and seal with nail polish. Image in fluorescence microscope using appropriate filters.

3.3. Quantitative Measurement of EGF–QD Binding and Internalization Using Flow Cytometry

Many properties of QDs, including their brightness and extreme photostability, make them very valuable for time-lapse fluorescence microscopy. However, a QD-labeled ligand or antibody is increased in size by 10–20 nm per bound QD. Therefore, carefully planned control experiments are necessary to establish that the QD does not interfere with the normal biological function of the ligand or antibody. Flow cytometry is an indispensable method for quantitation of binding and internalization data with high statistical precision. The following protocol can be used to show specific and saturable binding of EGF–QD to erbB1 on the cell surface, and demonstrate its uncompromised biological activity by monitoring internalization of the acceptor–ligand complex.

3.3.1. EGF–QD Specifically Binds to a Saturable Binding Site in the Membrane of erbB1-Expressing Cells

1. Prepare preformed EGF–QD at a ratio of 1:1 as described in **Subheadings 3.1.2.–3.1.3.**
2. Trypsinize A431 cells and resuspended in PBS/BSA at a density of 5×10^6/mL, and store on ice (*see* **Note 4**).
3. Prepare a twofold dilution series of EGF–QDs: make 1000 µL of the highest concentration of EGF–QD (100 n*M*) in PBS/BSA in a flow cytometry tube (*see* **Note 5**). Thorough mixing is very important. The tubes are kept on ice during the entire labeling procedure. Five hundred microliter is pipetted from this solution to another tube to which 500 µL of PBS/BSA is added yielding the second highest concentration (50 n*M*) of EGF–QD. This procedure is repeated until the lowest concentration is reached. The result is a dilution series of tubes filled with 500 µL of different concentrations of the ligand on ice (*see* **Note 6**).
4. Add 20 µL of the cell suspension to each ligand solution and mix thoroughly (*see* **Note 7**). Incubate the cells on ice for 60 min with regular mixing.
5. Measure the samples on a flow cytometer without washing (*see* **Note 8**). An unlabeled cell sample is also included in the measurement, the fluorescence intensity of which is subtracted from the mean fluorescence intensities of each sample. The corrected fluorescence intensities are plotted as a function of ligand concentration, and the characteristics of ligand binding are analyzed.

3.3.2. Binding of Streptavidin-Coated QDs to Cells Labeled With Different Molar Ratios of EGF and Biotin-EGF

Results of the previous experiments showed that EGF–QD binds to a saturable binding site on the surface of A431 cells. Strong evidence for specific binding is provided by competitive inhibition by another ligand. Activation of erbB1 requires the binding of two EGF molecules to a dimer of the erbB1 receptor *(2,11)*. The relatively large size of QDs requires one to show unambiguously that the ligand–QD is functional. Furthermore, if erbB1 molecules are at high density on the cell surface of A431 cells, a bound QD may sterically hinder binding to a neighboring receptor. To rule out size or steric hindrance by the QDs, we labeled cells with different molar ratios of EGF and biotinylated-EGF, followed by SAvQD, and

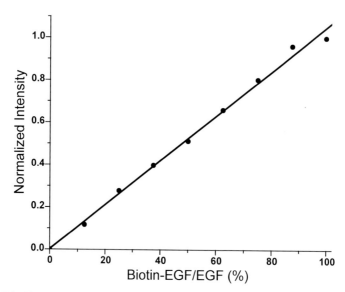

Fig. 3. Binding of streptavidin-cojugated quantum dots (SAvQDs) to cells labeled with different molar ratios of biotinylated- and non-biotinylated-EGF. A431 cells were incubated in the presence of different molar ratios of biotin-EGF and EGF for 60 min followed by labeling with SAvQDs. The fluorescence intensity of the samples (empty circles) was measured using flow cytometry. The correlation coefficient of a linear fit to the data (straight line) was R = 0.99. (Reproduced with permission from **ref. 8.**)

determined whether the fluorescence intensities of the samples were linearly proportional to the molar fraction of biotin-EGF. The higher the molar fraction of biotinylated-EGF, the higher the probability of finding two biotinylated-EGF molecules close to each other, e.g., binding to two erbB1 proteins in a dimer. This would result in leveling off of the fluorescence intensity curve at high biotin–EGF concentrations. The fact that we found a strictly linear relationship implies that binding of streptavidin-coated QDs is not sterically limited (*see* **Fig. 3**).

1. Trypsinize A431 cells and resuspend in PBS/BSA at a density of 2×10^6/mL. Store on ice (*see* **Note 4**).
2. Prepare 100 nM stock solutions of biotin–EGF and EGF in PBS/BSA. Keep the solutions on ice.
3. Prepare a 20 nM stock solution of SAvQD in PBS/BSA and store on ice.
4. Prepare a series of samples (total volume of each is 50 μL) in which the molar ratio of biotin–EGF changes from 0 to 100%, e.g. 0, 12.5, 25, 37.5, 50, 62.5, 75, 87.5, 100%). The total concentration of EGF (biotinylated + nonbiotinylated) is 100 nM in each sample.
5. Add 50 μL of cell suspension to each tube and mix thoroughly. Incubate the samples on ice for 30 min with regular shaking.

6. Layer 1 mL of fetal calf serum under the cell suspension and centrifuge the cells at 4°C. Discard the supernatant.
7. Add 300 µL of the 20 nM QD solution to each pellet and incubate the samples on ice for 30 min with regular shaking (*see* **Note 9**).
8. Without washing, measure the sample intensities using a flow cytometer.

3.3.3. Quantitative Measurement of EGF–QD Internalization

Although microscopy experiments can measure the internalization of ligand–QD in a qualitative or semi-quantitative manner with subcellular resolution, flow cytometry is indispensable for obtaining a quantitative description of the process.

1. Couple biotin–EGF to SAvQDs at a molar ratio of 6:1 as described in **Subheading 3.1.2.**
2. Trypsinize A431 cells and resuspended at a density of 10^6/mL in Tyrode's buffer plus.
3. Add 1 nM EGF–QD to the sample and incubate at 37°C with regular shaking.
4. Take 600-µL samples at 0, 5, 10, 20, and 30 min.
5. Add 5 mL of acid wash buffer to half of the sample, and incubate on ice for 5 min.
6. Wash with a large excess of PBS and resuspend cells in 300 µL Tyrode's plus.
7. Measure the fluorescence intensities of the acid washed sample and the non-acid-treated sample by flow cytometry to obtain the fraction of internalized ligand (*see* **Note 10**).

3.4. Other Types of In Vivo Labeling of Cells With Quantum Dots

QDs with many different types of conjugation are commercially available. Additionally, manufacturers have developed kits for alternative methods of bio-coupling QDs. We recommend the following recent papers and reviews for additional cell labeling techniques with quantum dots (*12–15*).

4. Notes

1. Thoroughly mix the SAvQDs and ligands by pipetting up and down several times. Mix in equal volumes so that the stoichiometry does not vary throughout the solution at the time of mixing.
2. Dilute the ligand–SAvQD complex from PBS into Tyrode's plus just before use because the SAvQDs aggregate with time in buffers containing divalent cations.
3. In performing similar experiments on another system than the EGFR, the researcher should adjust the concentration of ligand and incubation times according to the binding constant of the receptor or protein–ligand interaction.
4. The cell membrane is in a constant state of recycling. Antibodies and ligands typically induce the internalization of their receptor, and newly synthesized receptors continuously reach the plasma membrane by exocytosis. If cells are labeled at a temperature permissive to endo- and exocytosis, cells engulf EGF–QD

receptors, and the freshly exocytosed receptors will also bind EGF–QD from the continued presence of the label in the extracellular space. This results in a higher level of EGF–QD binding than would be expected based on the original number of receptors on the cell surface, giving rise to binding that is difficult to saturate. Therefore, both the cell suspension and the ligand dilution series are kept on ice during the whole duration of the experiment, and all solutions are prepared in flow cytometry tubes to minimize the number of times a solution has to be transferred from one tube to another. Although flow cytometers are usually not thermostated, if the tubes are kept on ice until measurement, ligand internalization is minimized.

5. The amount of receptor-bound ligand is usually negligible compared with the total amount. However, if a high-affinity ligand is investigated and the labeling volume is minimized, the decreased concentration of free ligand in the extracellular space (called ligand depletion) cannot be overlooked. For example, if 10^5 cells expressing 10^6 receptors/cell are labeled in the presence of 10 nM ligand in a volume of 500 μL, the number of bound ligands is $10^5 \times 10^6 = 10^{11}$ assuming saturation. The total number of added ligands is $10 \times 10^{-9}\ M \cdot 500 \times 10^{-6}\ \mathrm{L} \times 6 \times 10^{23}\ \mathrm{1/mol} = 3 \times 10^{12}$. In this case 33% of the total ligand amount is bound, severely distorting the calculations. As a rule of thumb, 5% ligand depletion is usually considered negligible. Beyond this limit either the real free-ligand concentration has to be determined, or corrections have to be introduced in the calculation, the details of which are beyond the scope of this chapter.

6. The ligand can also be depleted by binding to the wall of the labeling tube which can be minimized by carrying out the experiment in the presence of 0.1–0.5% (w/v) BSA. In addition, the same number of cells are labeled in identical volumes to ensure similar conditions (e.g., a constant number of total binding sites).

7. Dilution of the ligand after adding 20 μL cell suspension to 500 μL ligand solution is either neglected or can be corrected for in the following way: $c_{corr} = c_{orig} \cdot 500/520$, where c_{corr} and c_{orig} are the concentrations of the ligand in the diluted and undiluted solutions, respectively.

8. Ligand dissociation starts immediately when the concentration of ligand drops in the extracellular solution. Therefore, we suggest that samples not be washed before labeling. In the case of high-affinity ligands saturating their binding site at a low concentration, fluorescence of the extracellular ligands is negligible. In addition, flow cytometers are equipped with a constant background subtraction algorithm, reliably accounting for the contribution of extracellular fluorescence even in the case of low-affinity ligands.

9. The number of erbB1 proteins in 10^5 labeled A431 cells is $10^5 \times 2 \times 10^6 = 2 \times 10^{11}$ corresponding to $2 \times 10^{11} / 6 \times 10^{23} / 300$ μL = 1.1 nM mean receptor concentration. The concentration of QD was chosen to be 20 nM to ensure an approx 20-fold excess of unbound QDs.

10. Because acid washing removes non-covalently-bound molecules from the cell surface, the fluorescence intensity of acid-washed samples will represent the internalized pool of QD-labeled ligands.

Acknowledgments

This work was funded by EU FP5 Grants QLRT-2000-02278 and QLG1-CT-2000-01260 awarded to T.M.J and the Max Planck Gesellschaft.

References

1. Yarden, Y. and Sliwkowski, M. X. (2001) Untangling the ErbB signalling network. *Nat. Rev. Mol. Cell Biol.* **2**, 127–137.
2. Schlessinger, J. (2002) Ligand-induced, receptor-mediated dimerization and activation of EGF receptor. *Cell* **110**, 669–672.
3. Gadella Jr., T. W. J. and Jovin, T. M. (1995) Oligomerization of epidermal growth factor receptors on A431 cells studied by time-resolved fluorescence imaging microscopy. A stereochemical model for tyrosine kinase receptor activation. *J. Cell Biol.* **129**, 1543–1558.
4. Moriki, T., Maruyama, H., and Maruyama, I. N. (2001) Activation of preformed EGF receptor dimers by ligand-induced rotation of the transmembrane domain. *J. Mol. Biol.* **311**, 1011–1026.
5. Sako, Y., Minoghchi, S., and Yanagida, T. (2000) Single-molecule imaging of EGFR signalling on the surface of living cells. *Nat. Cell Biol.* **2**, 168–172.
6. Jorissen, R. N. Walker, F., Pouliot, N., Garrett, T. P., Ward, C. W., and Burgess, A. W. (2003) Epidermal growth factor receptor: mechanisms of activation and signalling. *Exp. Cell Res.* **284**, 31–53.
7. Hynes, N. E., Horsch, K., Olayioye, M. A., and Badache, A. (2001) The ErbB receptor tyrosine family as signal integrators. *Endocr. Relat. Cancer* **8**, 151–159.
8. Lidke, D. S., Nagy, P., Heintzmann, R., et al. (2004) Quantum dot ligands provide new insights into erbB/HER receptor-mediated signal transduction. *Nat. Biotechnol.* **22**, 198–203.
9. Lidke, D. S., et al. (2005) Reaching out for signals: filopodia sense EGF and respond by directed retrograde transport of activated receptors. *J. Cell Biol.* **170**, 619–626.
10. Grecco, H., Lidke, K. A., Heintzmann, R., et al. (2004) Ensemble and single particle photophysical properties (two-photon excitation, anisotropy, fret, lifetime, spectral conversion) of commercial quantum dots in solution and in live cells. *Micros. Res. Tech.* **65**, 169–179.
11. Mattoon, D., Klein, P., Lemmon, M. A., Lax, I., and Schlessinger, J. (2004) The tethered configuration of the EGF receptor extracellular domain exerts only a limited control of receptor function. *Proc. Nat. Acad. Sci. USA* **101**, 923–928.
12. So, M. K., et al. (2006) Self-illuminating quantum dot conjugates for in vivo imaging. *Nat. Biotech.* **24**, 339–343.
13. Alivisatos, A. P., Gu, W. W., and Larabell, C. (2005) Quantum dots as cellular probes. *Ann. Rev. Biomed. Eng.* **7**, 55–76.
14. Gao, X. H., et al. (2005) In vivo molecular and cellular imaging with quantum dots. *Cur. Opin. Biotech.*, **16**, 63–72.
15. Smith, A. M., et al. (2006) Engineering luminescent quantum dots for In vivo molecular and cellular imaging. *Ann. Biomed. Eng.* **34**, 3–14.

7

Single Quantum Dot Tracking of Membrane Receptors

Cédric Bouzigues, Sabine Lévi, Antoine Triller, and Maxime Dahan

Summary

The dynamics of membrane proteins in living cells has become a major issue to understand important biological questions such as chemotaxis, synaptic regulation, or signal transduction. The advent of semi-conductor quantum dots (QDs) has opened new perspectives for the study of membrane properties because these new nanomaterials enable measurements at the single molecule level with high signal-to-noise ratio. Probes used until now indeed encounter significant limitations: organic fluorophores and fluorescent proteins rapidly photobleach, whereas gold particles and latex beads, although more stable, are bulky and usually stain only one protein per experiment. In comparison, QDs are bright and photostable fluorescent probes with a size on the order of 10 nm can be used with standard immunochemical methods. We present the experimental protocols and methods of analysis which we used to investigate the dynamics of individual GABA$_A$ receptors in the axonal growth cone of spinal neurons in culture. Single QD tracking is nevertheless a general method, suitable to study many transmembrane proteins.

Key Words: Quantum dots; single molecule tracking; axonal pathfinding; GABA$_A$ receptors; fluorescence microscopy; immunocytochemistry; chemotaxis.

1. Introduction

Semiconductor quantum dots (QDs) are a new class of inorganic fluorophores with a potential to revolutionize biological detection and imaging *(1)*. Thanks to their remarkable optical and colloidal properties, they allow ultrasensitive multicolor measurements both in vitro and in vivo. A novel and promising applications of QDs lies in the possibility of targeting and detecting single nanoparticles. When inserted in a complex cellular environment, individual QDs act as nanoscale reporters that provide information on the dynamics of biological processes at a truly molecular level. By enabling, for instance, the direct visualization of molecular motions, QDs will contribute to a better understanding

From: *Methods in Molecular Biology, vol. 374: Quantum Dots: Applications in Biology*
Edited by: M. P. Bruchez and C. Z. Hotz © Humana Press Inc., Totowa, NJ

of the mechanisms that control and regulate the spatial distribution of key proteins in living cells.

Axonal guidance is a biological process on which much can potentially be learnt by measuring the lateral dynamics of individual key membrane proteins. Axonal pathfinding has been widely studied during the last decade and these investigations have led to the identification of numerous guidance cues. The corresponding signal transduction pathways have been explored and the role of many cellular actors has been described either by genetic tools or by guidance assay measurements *(2)*. Many questions remain though to explain the remarkable accuracy and spatial regulation of axonal guidance. In particular, it is still not known how a cell can correctly detect small gradients of external signals. In terms of gradient sensing, the membrane organization of receptors is probably critical *(3)* because dynamic changes in the spatial distribution of receptors might provide a way to amplify the gradient detection. We have studied the lateral dynamics of γ-aminobutyric acid (GABA) receptors which, in addition to their role in inhibitory synaptic transmission, are known to be involved in guidance mechanisms *(4)*. Lateral diffusion indeed appears to be a way for neurons to regulate their activity, from axonal pathfinding to synaptic regulation *(5)*. Using QDs and single molecule imaging in live neurons *(6)*, we have analyzed the motions of individual receptors in the membrane of axonal growth cones and have characterized their dynamics.

2. Materials

2.1. Culture

1. Minimum essential medium (Sigma-Aldrich Corp., St. Louis, MO, *see* **Note 1**).
2. Neurobasal medium supplemented with B27, glutamine (Invitrogen, Carlsbad, CA) and 25 µL antibiotics.
3. L-15 medium (Invitrogen) supplemented (for a 30-mL final volume) with 15 µL antibiotics, 30 µL progesterone to 10 µg/mL and glucose to 3.6 µg/mL.
4. B27 (Invitrogen) ready to use.
5. Glutamine (Invitrogen) diluted in water to 200 mM.
6. Polyornithine (Sigma-Aldrich Corp).
7. Fetal veal serum (FVS) (Invitrogen).
8. Deoxyribonuclease (Sigma-Aldrich Corp.).
9. Bovine serum albumin (Sigma-Aldrich Corp.).
10. Antibiotics: 10.000 UI penicillin and 10.000 µg/mL streptomycin (Invitrogen).
11. Glucose (Sigma-Aldrich Corp.).
12. Vectashield (Vector Laboratories, Burlingame, CA).
13. 10X Phosphate-buffered saline (PBS): 80.06 g NaCl, 2.02 g KCl, 2.05 g KH_2PO_4, 23.3 g Na_2HPO_4 in 1 L of deionized water.
14. Incubation buffer: 10% orthoboric acid solution (10 g of orthoboric acid dissolved in a 100 mL water solution adjusted to pH 8.0 and filtered on 0.22-µm pores) and

a 10% bovine serum albumin solution are added to water in respective volume proportion of 1/40 and 1/5. In order to obtain a final osmolarity around 260 mOsm, compatible with live cells, the incubation buffer has to be supplemented with 15% in volume of 1.43 *M* sucrose.

15. 4% Paraformaldehyde (PFA): 1 g PFA is dissolved in 22.5 mL of magnetically agitated water at 80°C. If solution remains unclear after agitation, a droplet of 1 *M* NaOH is added. Solution is then completed with 2.5 mL 10X PBS and stored at –20°C.

2.2. Antibodies and QDs

1. Antibody against γ2 subunit of GABA$_A$ receptor raised in rabbit (Euromedex, Mundolsheim, France) at 0.6 mg/mL stored at –20°C in deionized water.
2. Fab fragment of biotinylated goat anti-rabbit (Jackson Immunoresearch Laboratories, West Grove, PA) stored at –20°C in 50–50 glycerol-water buffer at 0.5 mg/mL.
3. 2 µ*M* Streptavidin-coated QDs emitting at 605 nm (Qdot 605 ITK Streptavidin Conjugates [1000-1], Quantum Dot Corp., Hayward, CA) stored at 4°C.

2.3. Fluorescence Video Microscopy

1. The experiments are carried out using a standard inverted microscope (Olympus, IX71) mounted with an oil objective (Olympus ×60 N.A. = 1.4 PlanApo Ph3) (*see* **Notes 2** and **3**).
2. QDs are excited with a 100 W mercury lamp (excitation filter XF1074 525AF45, dichroic mirror XF2017 DRLP 560, emission filter XF3304 605WB20; Omega Optics, Brattelboro, VT). Fluorescence images are collected with a Coolsnap ES (Roper Scientific) or a Peltier-cooled Micromax EB512FT charge-coupled device camera (Roper Scientific, Trenton, NJ).
3. Circular glass cover slips (18 mm; Karl Hecht KG, Sondheim, Germany).

2.4. Data Processing

Acquisition of fluorescence image stacks is performed with MetaView (Universal Imaging, Downingtown, PA). Tracking and statistical analysis are carried out with a custommade software using MATLAB® (Mathworks, Natick, MA). Software will be made available at the following website: http://www.lkb.ens.fr/recherche/optetbio/.

3. Methods
3.1. Cells and Medium Preparation

1. Cover slips are coated with polyornithine for one night and for a few hour with L15 medium supplemented with fetal veal serum (5% in volume) and 7% NaHCO$_3$ (2.5% in volume) before cells plating. Spinal cords of E14 rat embryo

are dissected in PBS-glucose, incubated for 15 min at 37°C in 1 mL of PBS supplemented with 20 μL of 2.5% trypsine-EDTA, and smoothly dissociated in deoxyribonuclease-supplemented L15 medium. Spinal neurons are then plated on cover slips at concentrations between 5×10^4/mL and 2×10^5 /mL in B27-Neurobasal medium and stored in sterile boxes in an incubator at 37°C under 5% of CO_2 until use for QD staining.

2. In order to work with neurons in ambient air, it is necessary to use a specific medium, referenced here as Mem-Air. Mem powder is diluted in 800 mL water, which is then completed with 6 g glucose, 4.5 mL 7% $NaHCO_3$, 10 mL HEPES, and sterilized by filtration on 0.22-μm pores. Mem-Air is completed immediately before experiments with 3 mL deionized water, 1 mL B27, 500 μL sodium pyruvate, and 500 μL glutamine for a 50-mL final volume.

3.2. Calibration of the Optical System for Single Molecule Detection

To evaluate the ability of the optical system (objective, filter, camera) to detect individual QDs, a simple procedure can be followed.

1. Prepare a clean cover slip by successively rinsing it with methanol, acetone, and water. Dry it with clean compressed air or nitrogen.
2. Deposit a drop of QDs diluted at 1 nM in water. After 5–10 min, rinse with water and dry with clean compressed air or nitrogen.
3. Mount the cover slip on the microscope and focus on the surface. To facilitate the focusing, a mark from a fluorescent pencil can be added on the proper side of the cover slip.
4. The signal of individual QDs corresponds to diffraction-limited spots and should be easily observable with the camera or in the eyepieces. The spots have a spatial extension given by λ/N_A where $\lambda = 605$ nm is the emission wavelength and N_A the numerical aperture of the objective. Individual QDs are identified by their fluorescence intermittency, i.e., by the random succession of periods with high and low emission intensity.
5. If necessary, the blinking rate can be decreased by reducing the excitation intensity with a neutral density. As this will diminish the emission signal accordingly, a compromise should be found between detection sensitivity and blinking.

3.3. Labeling for Single Molecule Experiments

The concentrations of labeling reagents (primary and secondary antibodies, QDs) indicated next correspond to conditions we found appropriate for our experiments. For a different biological system, the immunolabeling conditions might vary depending on the antigens and the antibodies, and should be adjusted accordingly (*see also* **Notes 4** and **5**). When testing for the right dilution, the cells should be imaged in live conditions because fixation might induce an additional auto-fluorescent background which would complicate the detection of individual QDs.

1. Remove a cover slip from the culture boxes. Place it on a surface maintained at 37°C and incubate with 500 μL of MEM-Air.
2. Aspirate the medium and add 100 μL primary antibody (Anti-γ2, 1:100) on the cover slip for 10 min at 37°C.
3. Discard the primary antibody and wash the sample three (or more) times in 500 μL Mem-Air.
4. Add 100 μL of secondary antibody (biotinylated goat anti-rabbit, 1:200) for 10 min at 37°C.
5. Discard the secondary antibodies and rinse the sample three times in 500 μL Mem-Air.
6. Prepare the QDs at 1:2000 to 1:10,000 in the incubation buffer.
7. Incubate the cover slip for 1 min with QDs then wash at least five times in Mem-Air to remove free QD remaining in solution.

At this point, the cells can be visualized live on the microscope or fixed for further immunocytochemistery. In this case, the following protocol can be used.

1. Immediately after labeling, fix the neurons in 4% PFA for 15 min.
2. Wash quickly three times in PBS.
3. Incubate 15 min with 33 mM NH$_4$Cl in PBS.
4. Permeabilize cells with 0.25% Triton in PBS for 1 min.
5. Wash in 0.25% gelatin in PBS.
6. Prevent nonspecific staining by incubation for at least 25 min with 0.25% gelatin in PBS.
7. The cover slips are ready for antibody staining.
8. After staining, rinse the cover slips in water and mount them on a microscope slide using a mounting buffer such as Vectashield.
9. Slides are stored at –20°C before observation on a fluorescent microscope and can be conserved for several months.

3.4. Single Molecule Imaging on Live Cells

1. Cover slip is mounted on a chamber, maintained either at room temperature or at 37°C.
2. By means of a phase-contrast objective, select a zone of interest in a cell. This selection can be also achieved with less sensitivity with bright field imaging.
3. Turn off the transmission light and turn on the fluorescence excitation. Individual spots can be detected with the camera or in the eyepieces. If necessary, slightly refocus the microscope.
4. A sequence of images can then be acquired. In our experiments, we performed two different types of recordings: time lapse or streaming. In both cases, the excitation illumination is controlled by a computer-driven mechanical shutter. For streaming acquisition, the shutter is permanently open allowing a continuous recording at an acquisition rate of about 13 Hz (75 ms/frame). The maximum number of images in the sequence is determined by the storage space in the RAM or the hard drive of the computer. The acquisition rate is usually limited by the read-out time of the

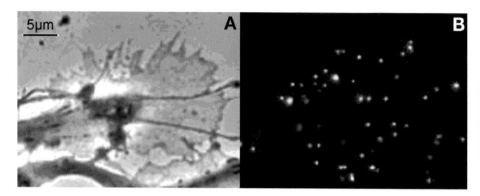

Fig. 1. Transmission image of a growth cone and fluorescence image of single quantum dot coupled to GABA$_A$R of the membrane.

camera and not by the brightness of the probe (static single QDs can be detected with a 5- or 10-ms acquisition). For time lapse acquisition, a 75-ms image was acquired every second and the shutter was opened and closed synchronously.

5. The sequences of fluorescence (*see* **Fig. 1**) contain fluorescence spots coming either from single QDs or from small aggregate. The blinking of QDs provides a simple criterion to identify single molecules, because the probability for all the QDs forming an aggregate to simultaneously blink is small.

3.5. Control of the Endocytosis of the Labeled Membrane Receptors

It is important to ensure that all tracked receptors are still in the membrane for the duration of experiments. In our experiments, this was checked by an acid wash *(7)* procedure: a quick passage in an acidic medium removes all membrane staining, whereas internalized receptors remain unaffected by the acid treatment.

1. The acid buffer is prepared from the original Mem-Air, whose pH is adjusted to 2.0 with a small quantity of 1 *M* HCl and kept at 4°C.
2. The labeled neurons are kept at room temperature or 37°C for a given period of time to let endocytosis occur.
3. The cover slip is placed on a cold surface at 4°C. It is rinsed and incubated with the acid buffer for less than 2 min.
4. The cover slip is abundantly washed with Mem-Air and ready to be mounted on the microscope to check the endocytosis. If necessary, the cells can be fixed with PFA for later observation.

For GABA$_A$R, it appeared that during the typical duration of an experiment (about 30 min), endocytosis could be neglected.

3.6. Analysis of Single QD Tracking Data

We describe the method of tracking that has been used on our data *(8)*.

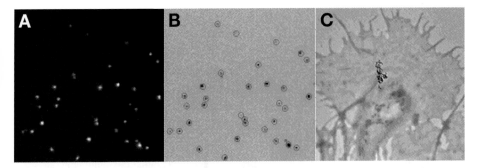

Fig. 2. (A) First image of a fluorescence movie. (B) Detection of the spots on the image and (C) reconstruction of one trajectory.

1. Measurement of the point-spread function (PSF). Our method for the detection and tracking of QDs requires the measurement of the PSF, which is the response of the optical system to a point source. QDs being much smaller than λ can be reliably considered as point source and an image of a single QD is a good approximation of PSF.

2. Detection of spots. Fluorescent spots are detected on each frame of the stack by a method of correlation. A measure of similarity between real fluorescent image and the PSF (known from calibration images) is computed around local maxima of a small region of the fluorescence image. Correlation maxima, selected above a threshold dependent on our estimation of the noise in the fluorescence image, reveals positions of QDs. (*see* **Fig. 2A–B.**)

3. Individual trajectories. Once QDs are detected on each frame, spots coming from the same QD have to be linked from frame to frame. Assuming that receptors are freely diffusing, we computed the probability for a spot, detected at a given position on a frame to be at another position on the following frame. The highest likelihood for a spot of the following frame then provides the correct reconstruction of the trajectory. Spots of a given frame are then associated to the trajectory they are the most likely to belong to. (*see* **Fig. 2C.**)

4. Blinking brings additional difficulties for the tracking of single QDs because the spots can temporarily disappear for random durations. A simple approach to account for blinking consists in computing continuous trajectories and then manually associating those corresponding to the same QD. To reduce the blinking rate, it is recommended to perform experiments with an excitation as low as possible.

3.7. Dynamics and Distribution of Receptors

1. Lateral dynamics of individual receptors. In order to study protein dynamics with a good time resolution, we recorded sequences of images composed of about 1000 frames (acquisition rate: 75 ms). These raw data are then processed to obtain a collection of receptor trajectories. Under conditions of labeling discussed in **Subheading 3.3.**, it was possible to obtain between 5 and 10 long trajectories

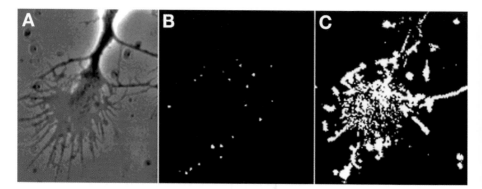

Fig. 3. (A) Transmission image of a growth cone. **(B)** Individual GABA receptors tagged with quantum dots. **(C)** Maximum of the fluorescence signal over the entire stack (800 images acquired at 1 Hz).

(composed of 500–1000 points) per sequence of images. These trajectories were then analyzed by computing the mean square displacement (MSD) (*see* **Subheading 3.8.**).

2. Cell surface explored by the receptors. Single QD imaging provides a simple way to visualize the fraction of the membrane explored by the tagged-proteins (**Fig. 3**). In this case, fluorescence measurements were performed in a time-lapse mode, with a 75-ms image aquired every second during 20–60 min. For each pixel in the image, the maximum signal in this pixel is computed over the entire image stack.

3.8. Statistical Analysis

One of the main issues of single particle-tracking experiments is to extract biological information from recorded trajectories. Here, we present some of the techniques that have been used in order to measure quantities, such as the diffusion coefficient or velocity.

3.8.1. MSD

A way to extract physical information from a trajectory *(x[t], y[t])* is the computation of *ρ(t)* the MSD (*9*). It is determined by the following formula:

$$\rho(n\tau) = \frac{1}{N-n}\sum_{i=1}^{N-n}\left[(x((i+n)\tau) - x(i\tau))^2 + (y((i+n)\tau) - y(i\tau))^2\right]$$

where τ is the acquisition time and N the total number of frames. Through computation of the MSD, the nature of the molecular motions can be analyzed. A diffusive motion is revealed by a MSD varying linearly with a slope *4D* (where *D* is the diffusion coefficient). In practice, the parameter *D* is obtained by fitting the first five points of the MSD with a linear curve. If an additional directed

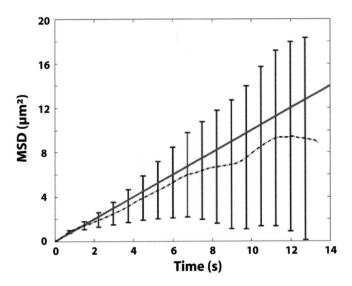

Fig. 4. Mean square displacement (dashed line) for a 30-s long simulated trajectory (D=0.25 μm²/s) with the corresponding error bars and the linear fit (plain line).

motion (with velocity **v**) is present, **ρ(t)** is in average equal to **4Dt+v²t²**. However, directed motions are more difficult to detect because the statistical errors in the MSD increase with time (*see* **Subheading 3.8.2.**).

3.8.2. Statistical Errors

Because the MSD is calculated on trajectories with a finite number of points, it is affected by statistical errors (**Fig. 4**). σ(τ) the variance of ρ(τ) can be either estimated by computer simulations or, in the case of Brownian diffusion, computed by the following formula *(10)*:

$$\sigma(\tau)=4D\tau\left[(4n^2(N-n)+2(N-n)+n-n^3)\,/\,6n(N-n)^2\right]^{1/2} \text{ for } n \leq N/2$$

$$\sigma(\tau) = 4D\tau\left[1+\left((N-n)^3 - 4n(N-n)^2 + 4n - (N-n)\right)/\,6n^2(N-n)\right]^{1/2} \text{ for } n \geq N/2$$

3.8.3. Detection of Directed Motions

In order to detect directed motions, it is necessary to compute the MSD with good statistical accuracy. This can be achieved by two different methods: (1) averaging the MSD curves calculated for different trajectories assuming a homogenous population of proteins, or (2) recording long enough trajectories (typically 100 s for the range of speed of drift we have been able to measure). This emphasizes the interest of using QDs instead of organic dyes, which do not allow such measurements because of their rapid photobleaching.

4. Notes

1. All solutions should be prepared in water with a resistivity of 18.2 MΩ.cm and a total organic content of less than five part per billion. In our laboratory, water was obtained from a Milli-Q system from Millipore.

2. For single molecule imaging, it is essential to use an oil- or water-immersion objective of good quality with high numerical aperture (at least 1.2 and preferably 1.3–1.45). Single QDs should be detected with an epifluorescence microscope and not a scanning (confocal or two-photon) microscope.

3. The pixel size, p, of the camera and the magnification, M, of the objective should be such that p/M is no larger than $\lambda/2N_A$. In our case, $p = 13$ μm and $M = 60$ such that $p/M = 217$ nm. If p/M is too small, each individual spot will spread over too many pixels leading to a decrease in the signal-to-noise ratio.

4. It is important to choose a concentration with little nonspecific adhesion. In all our experiments we found the suitable dilution of QDs to be between 1:1000 and 1:10,000 according to antibodies. Concentration of primary antibody is usually close to those used in classical immunocytochemistry, and concentration of secondary antibody has to be chosen as the minimal concentration giving a satisfying staining. It is, in fact, not particularly critical when using Fab fragments because the probability of cross-linking the primary antibodies is greatly reduced. To even further decrease this probability, it is possible after labeling to saturate the free binding sites of the streptavidin-coated QDs with an excess of biotin. However, we never observed that it resulted in a significant difference in the diffusion properties of the tagged proteins.

5. It has become possible to directly conjugate QDs to primary antibodies through covalent binding. For more information, check the protocols on the web site of Quantum Dot Corp. (www.qdots.com) or Evident Technology (http://www.evidenttech.com/).

Acknowledgments

We are grateful to Marie-Virginie Ehrensperger, Camilla Luccardini, Béatrice Riveau, Philippe Rostaing, Stéphane Bonneau, and Laurent Cohen for their help and contribution.

References

1. Michalet, X., Pinaud F. F., Bentolila, L. A., et al. (2005) Quantum dots for live cells, in vivo imaging, and diagnosis *Science* **307**, 538–544.

2. Song, H.-J. and Poo, M. M. (2001) The cell biology of neuronal navigation. *Nature Cell Biol.* **3**, E81–E88.

3. Guirland, C., Suzuki, S., Kojima, M., Lu, B., and Zheng, J. Q. (2004) Lipid rafts mediate chemotropic guidance of nerve growth cones. *Neuron.* **42**, 51–62.

4. Xiang, Y., Li, Y., Zhang, Z., et al. (2002) Nerve growth cone guidance mediated by G protein-coupled receptors. *Nat. Neurosci.* **5**, 843–848.

5. Choquet, D. and Triller, A. (2003), The role of receptor diffusion in the organization of the postsynaptic membrane. *Nat. Rev. Neurosci.* **4**, 251–265.

6. Dahan, M., Lévi, S., Luccardini, C., Rostaing, P., Riveau, B., and Triller, A. (2003) Diffusion dynamics of glycine receptors revealed by single quantum dot tracking. *Science* **302,** 442–445.

7. Tardin, C., Cognet, L., Bats, C., Lounis, B., and Choquet D. (2003) Direct imaging of lateral movements of AMPA receptors inside synapses. *EMBO J.* **22,** 4656–4665.

8. Bonneau, S., Cohen L., and Dahan, M. (2004) A multiple target approach for single quantum dot tracking. *Proceedings of the IEEE International Symposium on Biological Imaging* (ISBI 2004), p. 664.

9. Saxton, M. J. and Jacobson, K. (1997) Single-particle tracking: applications to membrane dynamics. *Annu. Rev. Biophys. Biomol. Struct..* **26,** 373–399.

10. Qian, H., Sheetz, M. P., and Elson, E. L. (1991) Analysis of diffusion and flow in two-dimensional systems. *Biophys. J.* **60,** 910–921.

8

Optical Monitoring of Single Cells Using Quantum Dots

Jyoti K. Jaiswal and Sanford M. Simon

Summary

Quantum dots (QDs) can be used to label live cells for long-term tracking experiments. The stability of the QDs ensures low toxicity and long-term imaging capability. The multicolor capabilities of QDs allows the simultaneous monitoring of complex cellular populations, allowing more information-rich experiments to be performed. This chapter discusses the ways in which one can label cells with QDs, and a number of ways to characterize the impact that the labeling has on the cellular populations.

Key Words: Cellular labeling; tracking; migration; metastasis; differentiation; quantum dot; photostable; microinjection; transfection.

1. Introduction

To decipher how multicellular organisms achieve and maintain their complexity, it is not only necessary to monitor multicellular and multimolecular interactions in vivo, but to monitor them at high resolution and for periods that may extend over several days. A variety of tools are being developed to help achieve these goals. These tools include novel imaging modalities such as multiphoton microscopy and a growing list of organic fluorophores such as fluorescent dyes and genetically encoded fluorophores *(1,2)*.

Recently inorganic fluorophores have been added to the list of fluorescence-based reporters for biological imaging. Fluorescent semi-conductor nanocrytals, known as quantum dots (QDs), are well suited for simultaneous imaging of multiple biological samples for extended periods *(3)*. This is because of their high photostability, broad excitation and narrow emission spectra, large Stoke's shift, ability to achieve specific in vivo labeling, and absence of detectable deleterious effects. In this chapter, we will describe the use of QDs for labeling live cells and tracking individual cells in culture and in vivo. We will also describe

From: *Methods in Molecular Biology, vol. 374: Quantum Dots: Applications in Biology*
Edited by: M. P. Bruchez and C. Z. Hotz © Humana Press Inc., Totowa, NJ

the use of spectral microscopy for multicolor imaging for monitoring multiple populations of QD-tagged cells.

2. Materials

1. Hydrophilic QDs. These experiments require water-soluble QD. The QDs can be synthesized and conjugated to specific biomolecules in the investigator's laboratory as described elsewhere *(4)* or purchased from companies such as Quantum Dot Corporation (Hayward, CA) or Evident Technology (Troy, NY). These suppliers also provide QDs conjugated to avidin or biotin or to specific antibodies. Here, we describe the use of CdSe/ZnS QDs rendered water soluble by capping with dihydroxylipoic acid (DHLA) *(5)*.
2. Cells and organisms of interest (prokaryotic or eukaryotic organisms and cells can be used).
3. Fluorescence or laser scanning confocal microscope. Any microscope that is equipped with an excitation light source and appropriate optical filters for fluorescence can be used. To resolve multiple variants of QDs simultaneously on a fluorescence microscope, their emission could be spectrally resolved by using a fluorescence spectra scanning device such as Spectral-Vue from Optical Insights (Tucson, AZ) or a PARISS system from Light Form Inc., (Hillsborough, NJ). Alternatively, a filter-based setup such as Dual-View or Quad-View splitter from Optical Insights can be used. For confocal imaging Zeiss META detector (Jena, Germany) or the AOBS module for the Leica (Heidelberg, Germany) laser scanning confocal microscope may be used. The large Stokes', shift for the fluorescence of QDs makes it easier to image multiple fluorophores as only a single wavelength of excitation light is required for QDs of all colors. Commercially available QDs emit all the way from blue to near infra red (NIR) region of the spectrum. Thus the microscope should be equipped with filters depending on the emission spectrum of the quantum dot to be used.
4. Phosphate-buffered saline (PBS; Sigma Chemicals).
5. Sulfo-NHS-SS biotin (Pierce Biotechnology).
6. Tris-buffered saline (Sigma Chemicals).
7. Lipid-based transfection reagent (e.g., Lipofectamine 2000 [Invitrogen] or Fugene 6 [Roche Diagnostics, Indianapolis, IN]).

3. Methods

The methods described next outline the approaches for labeling live cells with QDs, testing for effects of labeling with QDs on cell physiology, and imaging of single or multiple populations of QD-labeled cells in vivo and in vitro.

3.1. Labeling

For tracking cells, QDs can be used to label either the extracellular surface or loaded inside of cells *(6–10)*. For extracellular labeling of live cells, QDs can be conjugated to ligands or antibodies that bind the proteins present on the surface of the cell. This approach, described next, has been used for both in vitro and in vivo labeling of live cells. A number of approaches can be used for labeling the inside of cells. Because cell surface proteins are endocytozed, much of the cell surface label gets internalized over time. Cells can also be labeled nonspecifically by incubating them with QD, which will be gradually internalized by endocytosis. Alternatively, the QD can be introduced directly into the cytosol by a number of techniques and reagents, including microinjection, amphipathic peptides, or lipids.

3.1.1. Endocytic Labeling

This is a relatively noninvasive approach for intracellular labeling of cells *(6)*.

1. Incubate cells in growth media with up to 2 µM QDs for 2–3 h.
2. Remove the extracellular QDs by washing the cells several times with growth media or a physiological buffer.

3.1.2. Labeling with Amphipathic Peptides

Use of a carrier peptide accelerates labeling of cells and requires fewer QDs. However, it is still dependent on the ability of cells to endocytose *(11)*. For intracellular delivery of QDs into live cells, Pep1 peptide can be used *(11,12)*.

1. Prepare the Pep1–QD complex by mixing 10 µM peptide and 100–500 nM QDs in 100 µL of serum-free media.
2. Add this mixture to 2 mL of serum-free media, and incubate cells in this media for 1 h.
3. Prior to imaging the cells, remove the extracellular QDs by washing with serum-free media or a physiological buffer.

3.1.3. Labeling with Cationic Lipids

This approach also allows efficient and rapid intracellular delivery of QDs *(7)*.

1. Prepare lipid–QD complex by mixing 100 µL of serum-free medium with 100 pmole of negatively charged QDs (e.g., capped with DHLA or other carboxylate-containing reagents) QD and 4 µL of lipofectamine-2000 or 3 µL of Fugene-6.
2. Add the mixture from **step 1** to cells in media with or without serum and incubate for 1–2 h.
3. Wash the cells free of extracellular lipid–QD conjugate.

3.1.4. Labeling by Microinjection

This approach can be used to for targeted intracellular delivery of QDs into cells in culture or in vivo *(13,14)* (**Fig. 1**).

1. Dilute QDs to 10–100 nM in buffer containing 140 mM KOAc and 10 mM HEPES (pH 7.4) or another buffer suitable for injecting into cells.
2. Inject desired amount of QD solution into the cells in culture or in vivo.

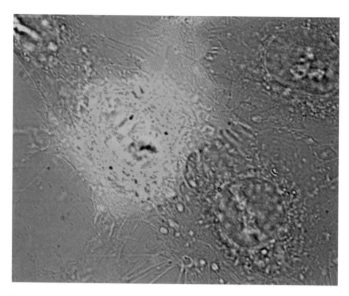

Fig. 1. Quantum dot (QD)-labeling of mammalian cells by microinjection. HeLa cells growing on a cover slip were injected with 50 n*M* solution of polyethylene glycol-coated QDs *(14)* suspended in the injection buffer described in the text (*see* **Subheading 3.1.4.**). The cells were imaged using epifluorescence microscopy with 488-nm excitation light and 63X 1.45NA oil immersion objective. The QD fluorescence is pseudocolored green and overlayed on the bright field image of the same region.

3.1.5. Biotin-Mediated Cell Surface Labeling

This approach involves biotinylating the cell surface molecules and then allowing binding of avidin-conjugated QDs *(6)*. This can be used to label either prokaryotic or eukaryotic cells. However, biotinylation of cell surface molecules could affect the cell surface properties, such as adhesion, cell–cell interaction, and signaling.

1. Wash cells free of growth medium using PBS or an appropriate buffer.
2. Incubate cells in a freshly prepared solution of 1 mg/mL Sulfo-NHS-SS biotin in the buffer of choice.
3. Incubate cells either for 30 min at 4°C or for 5 min at room temperature, and then quench the excess biotin by washing the preparation with Tris-buffered saline (pH 7.4).
4. Incubate the cells for 10 min in a serum-free medium containing 0.5–1 n*M* avidin-conjugated QDs.
5. Prior to imaging, remove the unbound QDs by repeated washing with the buffer.

3.1.6. Targeted Labeling

QDs conjugated to a specific antibody (Ab), a ligand, or to avidin/sterptavidin (SAv) are needed *(6,9,10)* (**Fig. 2**). For targeted labeling of cells in vitro, the

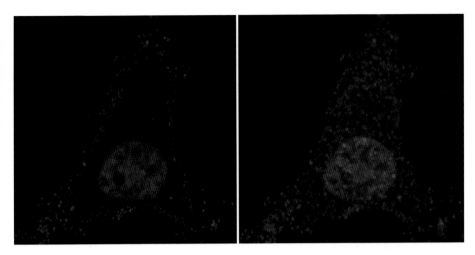

Fig. 2. Targeted labeling of transferin receptors on the HeLa cell surface using quantum dots (QDs). HeLa cells growing on the cover slip were incubated for 30 min in serum-free media and then for 15 min at 4°C with serum-free media containing 50 mg/mL of biotinylated transferrin. The cells were washed three times with cold PBS and incubated for 10 min with 565-nm emitting QD conjugated with streptavidin. Cells were fixed with 4% paraformaldehyde and stained with Hoechst-3342 to mark the nuclei before confocal imaging, using 60X 1.2NA water immersion objectives. The image on the left shows a 1-μm thick optical section of a labeled cell, whereas that on the right shows the three-dimensional projection of the entire cell. QD fluorescence is psuedocolored red and the nuclear dye fluorescence is psuedocolored blue.

concentration of the QD bioconjugate is determined by the affinity of the Ab for the target antigen. For labeling cells in vivo, additional factors such as efficiency of delivery and the rate of clearance of the QDs from the body causes additional variability. Thus, the exact QD bioconjugate concentration to be used must be determined in each case.

1. For in vitro labeling, incubate cells with QD–Ab or QD–ligand bioconjugate, or with appropriate biotinylated antibody for 20 min.
2. Wash the cells twice with appropriate media or buffer to remove excess unbound material.
3. When labeling with QD–SAv bioconjugate, following the incubation with biotinylated antibody incubate cells for another 10 min with the QD–SAv bioconjugate, followed by two washes.
4. For targeted labeling in vivo, administer an appropriate amount of QD bioconjugate to the animal. Injecting 0.4 nmol of antibody-conjugated QDs through the tail vein have been shown to provide successful labeling of tumor cells in vivo (**8**).
5. Carry out the imaging after the unbound QD bioconjugates have been cleared from the circulation.

3.1.7. Multicolor Labeling

To label one cell type or populations of cells with different colored QDs, either the lipid-based or carrier peptide-based strategy could be used *(12)*.

1. Mix in an equal quantity of QDs of the desired colors and then proceed with the peptide or cationic lipid-based labeling approach as described previously.
2. When labeling different cells with a combination of different color QDs, be aware that neither of these approaches will deliver exactly equal numbers of QDs of each color in every cell. The greater the variety of QDs, the greater will be the variability. Thus, it is necessary to quantify the relative fluorescence of each color QDs in more than100 randomly chosen cells.
3. If greater uniformity of labeling is desired, then a fluorescent cell sorter can be used to sort the labeled cells according to the desired uniformity of labeling.

3.2. Testing for Effects of Labeling

The use of QDs in live cells and animals does not lead to cytotoxicity *(6,7,9,15)*. However, some reports have suggested that in some circumstances the nature of the core and surface coating of QDs can affect cell physiology *(16,17)*. Thus, before using the QD as a label for a particular cell type, it is important to determine if that particular formulation of QD has any detectable effects of the physiology of the cells. Depending on the exact application for which the labeled cells will be used, assays should be performed to compare labeled cells with the unlabeled cells. Here, we will illustrate this by taking cell proliferation, cell motility, and cell signaling as examples. In each example, cells labeled with QD are compared side-by-side with cells that have not been labeled, but are otherwise indistinguishable.

3.2.1. Cell Proliferation

1. Plate a mixture of labeled and unlabeled cells on a gridded cover slip.
2. Three to four hours after plating, count the number of labeled and unlabeled cells in 8–10 squares from a different part of the cover slip.
3. Repeat the counts at least three times every 12–48 h depending on the doubling time of the cells.
4. Compare the rate of doubling of labeled vs the unlabeled cells in the same cover slip.

3.2.2. Cell Motility

1. For cells that show chemotaxis (e.g., neutrophils, *Dictyostelium*, and so on) put the labeled and unlabeled cells in a chemotactic gradient (set up by placing a micropipet with the chemotactic substance some distance away from the cell mass) and monitor the speed and direction of cell movement in this gradient.
2. For nonchemotactic cells, plate a mixture of QD labeled and unlabeled cells at >90% confluence and let the cells attach for 6 h.

3. Scratch the confluent monolayer using a micropipet tip to create a cell-free zone.
4. Count the number of labeled and unlabeled cells lining this cell-free zone.
5. Incubate the cells 3 h under optimal growth conditions and then determine what fraction of labeled and unlabeled cells that were lining the cell-free zone have migrated into it.
6. An alternative formulation is to mix cells labeled with QD and cells labeled with a cell tracker dye, plate them at >90% confluence, and let the cells attach for 6 h.
7. Quantify the ratio of the fluorescence of cell tracker dye to QD in a few fields.
8. Scratch the confluent monolayer as in **step 3**.
9. Quantify the velocity at which the QD and cell tracker-labeled cells fluorescence move into the cell-free zone.
10. Incubate the cells 3 h under optimal growth conditions and repeat the measurement. Normalize the QD and cell tracker fluorescence in the cell-free zone to the fluorescence recorded in **step 7**.

3.2.3. Cell Signaling and Differentiation

Cell differentiation is another test of the effect of QD labeling on the cellular physiology. Most cultured animal cells are terminally differentiated precluding their use for such an analysis. Thus, undifferentiated cells lines (e.g., the muscle precursors, the hepatocyte precursor, stem cells, and so on) should be used to monitor the effects of different QDs and QD bioconjugates on cell differentiation. Irrespective of the cell line and the means used to trigger differentiation, the approach to evaluate the effect of QD-labeling on cell differentiation would be the same. Here, this approach is described using the cellular slime mold *Dictyostelium discoideum*.

1. Label cells using any of the approaches described in **Subheading 3.1**.
2. Mix these cells with unlabeled cells and plate them together.
3. Trigger differentiation using the appropriate treatment. In *Dictyostelium* cells starvation by removal of growth media triggers differentiation.
4. To determine any major effects on cell differentiation, monitor the labeled cells for obvious morphological changes such as elongated shape and pulsatile chemotaxis resulting in the formation of an aggregate.
5. Monitor minor and long-term effects by allowing the organism to complete development and following the fate of labeled and unlabeled cells in the developing organism.
6. Independently test for the kinetics and level of expression of key protein markers for the progression of development by sacrificing the developing organism and immunodetection of these markers in the QD-labeled versus unlabeled cells.

3.2.4. Effect of QD-Labeling on Cells In Vivo

The ability of cells to form tumors in vivo requires the ability to trigger expression of a variety of genes that would allow surviving the mechanical stress, chemotaxing, growing in vivo, and being able to induce the formation of blood vessels. Thus to assess the effect of QD labeling on cell in vivo, the ability of labeled cells to form tumors in vivo provides an effective assay and can be carried out as follows *(7)*:

1. Using approaches described in **Subheading 3.1.** label tumor cells with QDs.
2. Mix these with cells labeled with an independent fluorescent maker (a cell tracker dye or cells expressing green fluorescent protein [GFP]). Inject the mixed population of cells in a pair of healthy syngenic, age- and sex-matched mice.
3. Allow for the tumors to form and grow for an appropriate period.
4. Extract the target organ and compare the number and size of tumors formed by the labeled and unlabeled cells.

3.3. Imaging

QDs can be imaged using all types of fluorescence microscopes and using lamp- or laser-based excitation source. Unlike organic fluorophores, which can only be excited by a narrow region of wavelengths near their emission, the QD can be excited by a broad-band of excitation. Indeed, the efficacy of the excitation increases as the excitation wavelength is shifted to shorter wavelengths away from the peak emission. Thus they are better excited by ultraviolet (UV) light. Because the UV light causes greater damage to live cells, it is recommended that the wavelength of the excitation light be kept above 400 nm. Fortunately, QD are easily excited with multiphoton light. Excitation at 800 nm is optimal, but any wavelength of light between 700 and 1000 nm could be used *(18)*. This has multiple advantages including eliminating UV-induced photodamage and facilitating excitation of QD that may be deep within biological samples.

3.3.1. Microscope Setup

The filter sets for use with the commercially available QDs can be purchased from Chroma Technology Corporation (Rockingham, VT). The rest of the microscope setup should be the same as for the imaging conventional fluorescent dyes. When cells are being imaged for long periods care must be taken to minimize the effect of change in temperature or CO_2 level and constant illumination of sample on the health of cells.

3.3.2. Multicolor Imaging

For a mixture of cells labeled with different color QDs, imaging of each color QD could be carried out sequentially or simultaneously (**Fig. 3**). The kinetics and sensitivity of the process being studied would decide which mode of imaging is the best. Simultaneous use of multiple organic fluorophores results in overlapping excitation and emission spectra (**Fig. 4A**). This results in cross-talk of emission from more than one fluorophore. QDs allow the use of a single

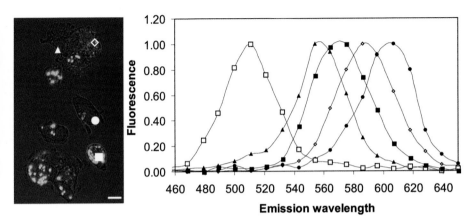

Fig. 3. Multicolor labeling and spectral imaging of mammalian cells. Different colored quantum dots were loaded into separate populations of the highly metastatic B16F10 melanoma cell line with lipofectamine 2000. The cells were imaged with a Zeiss multiphoton microscope with spectral resolution using a META detector. The figure is adapted from **ref. 7.**

excitation light to excite different color QDs and provide the ability to eliminate or minimize the overlap in emission spectra (**Fig. 4B**). To overcome cross-talk in the case of even minimal spectral overlap, spectral imaging (described in **Subheading 3.3.3.**) is the most effective approach. In case the filter-based approach is to be used for simultaneous or sequential imaging, it is important to correct for bleed-through of emission from one QD into the other QD channel.

3.3.3. Spectral Imaging Using LSM510 META Detector

Spectral imaging permits eliminating the effect of even significant overlap in emission spectra, thus improving the sensitivity of detection and enables improved imaging despite high background sample auto-fluorescence *(19)*. To carry out spectral imaging the following spectra should be obtained:

1. Auto-fluorescence spectra. Use an unstained cell or tissue sample to generate a reference spectrum for the tissue auto-fluorescence. Because auto-fluorescence changes with fixation protocol control spectrum should be obtained using the same fixation protocol.
2. QD fluorescence spectra. To obtain pure spectra of each color QD, mount each of the different color QD sample on the cover slip without any tissue or cell sample.

Excite both of the samples using the same excitation light and collect their emission spectrum using the META detector and store these emission spectra. Use these pure spectra for spectral identification of each color QD over the background auto-fluorescence. For further refinement, a spectral unmixing can be

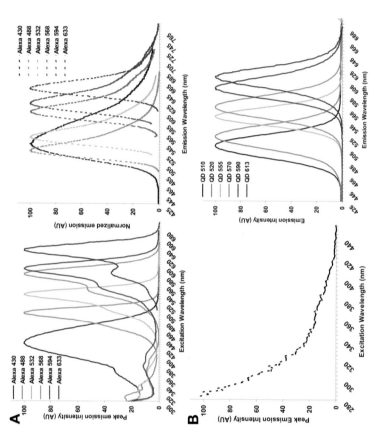

Fig. 4. (**A**) Excitation and emission spectra of Alexa fluor dyes. The values to plot the excitation and emission spectra for each of the different Alexa series dyes were obtained from the Invitrogen website (http://probes.invitrogen.com/servlets/spectra). The plot on the left shows excitation and that on the right shows emission spectra for each of the six dyes shown here. (**B**) Excitation and emission spectra of QDs. 1m*M* solutions of different color QDs (in borate buffer, pH 9.0) were excited using light of 350 nm and fluorescence emission recorded from 420 to 680 nm. Each spectrum was normalized to the peak emission. All the QDs show very similar excitation spectra, hence only one excitation spectrum is shown here. The excitation spectrum was obtained by recording emission at 570 nm of the QD 570.

performed following the data acquisition. By using the spectral imaging approach a high auto-fluorescent background from a tissue sample can be eliminated, converting it into a background-free image *(7)*.

3.4. Conclusion

This chapter focuses on the use of QDs for the sake of monitoring live cells. However, QDs are also useful for high-resolution imaging of proteins in live cells. The approaches to do so have been described elsewhere *(11,20)*. The lack of toxicity demonstrated for dihydroxylipoic acid (DHLA)-capped QDs and for some of the other commercially available QDs makes these fluorophores desirable for both single cell and single molecule imaging in live cells and organisms. It is nonetheless advisable that for each new organism and cell type and for every novel formulation of QDs, the effect of QDs on the viability and physiology of live cells must be evaluated. Careful use of these novel fluorophores holds the potential to enable us to address questions that have so far been elusive because of limitations of the organic fluorophores.

Acknowledgments

This work was supported by the grants NIH P20GM072015 and NSF BES 00520813 and NSF BES 0322867 to SMS.

References

1. Williams, R. M., Zipfel, W. R., and Webb, W. W. (2001) Multiphoton microscopy in biological research. *Curr. Opin. Chem. Biol.* **5,** 603–608.
2. Zhang, J., Campbell, R. E., Ting, A. Y., and Tsien, R. Y. (2002) Creating new fluorescent probes for cell biology. *Nat. Rev. Mol. Cell Biol.* **3,** 906–918.
3. Jaiswal, J. K. and Simon, S. M. (2004) Potentials and pitfalls of fluorescent quantum dots for biological imaging. *Trends Cell Biol.* **14,** 497–504.
4. Dabbousi, B. O., Rodrigez-Viejo, J., Mikulec, F. V., et al. (1997) (CdSe)ZnS core-shell quantum dots: Synthesis and characterization of a size series of highly luminescent nanocrystallites. *J. Phys. Chem.* **101,** 9463–9475.
5. Mattoussi, H., Mauro, J. M., Goldman, E. R., et al. (2000) Self-assembly of CdSe-ZnS quantum dot bioconjugates using an engineered recombinant protein. *J. Am. Chem. Soc.* **122,** 12,142–12,150.
6. Jaiswal, J. K., Mattoussi, H., Mauro, J. M., and Simon, S. M. (2003) Long-term multiple color imaging of live cells using quantum dot bioconjugates. *Nat. Biotechnol.* **21,** 47–51.
7. Voura, E. B., Jaiswal, J. K., Mattoussi, H., and Simon, S. M. (2004) Tracking metastatic tumor cell extravasation with quantum dot nanocrystals and fluorescence emission-scanning microscopy. *Nat. Med.* **10,** 993–998.
8. Gao, X., Cui, Y., Levenson, R. M., Chung, L. W., and Nie, S. (2004) In vivo cancer targeting and imaging with semiconductor quantum dots. *Nat. Biotechnol.* **22,** 969–976.

9. Lidke, D. S., Nagy, P., Heintzmann, R., et al. (2004) Quantum dot ligands provide new insights into erbB/HER receptor-mediated signal transduction. *Nat. Biotechnol.* **22**, 198–203.

10. Dahan, M., Levi, S., Luccardini, C., Rostaing, P., Riveau, B., and Triller, A. (2003) Diffusion dynamics of glycine receptors revealed by single-quantum dot tracking. *Science* **302**, 442–445.

11. Jaiswal, J. K., Goldman, E. R., Mattoussi, H., and Simon, S. M. (2004) Use of quantum dots for live cell imaging. *Nat. Methods* **1**, 73–78.

12. Mattheakis, L. C., Dias, J. M., Choi, Y. J., et al. (2004) Optical coding of mammalian cells using semiconductor quantum dots. *Anal. Biochem.* **327**, 200–208.

13. Dubertret, B., Skourides, P., Norris, D. J., Noireaux, V., Brivanlou, A. H., and Libchaber, A. (2002) In vivo imaging of quantum dots encapsulated in phospholipid micelles. *Science* **298**, 1759–1762.

14. Uyeda, H. T., Medintz, I. L., Jaiswal, J. K., Simon, S. M., and Mattoussi, H. (2005) Synthesis of compact multidentate ligands to prepare stable hydrophilic quantum dot fluorophores. *J. Am. Chem. Soc.* **127**, 3870–3878.

15. Ballou, B., Lagerholm, B. C., Ernst, L. A., Bruchez, M. P., and Waggoner, A. S. (2004) Noninvasive imaging of quantum dots in mice. *Bioconjug. Chem.* **15**, 79–86.

16. Derfus, A. M., Chan, W. C. W., and Bhatia, S. N. (2004) Probing the cytotoxicity of semiconductor quantum dots. *Nano Letters* **4**, 11–18.

17. Shiohara, A., Hoshino, A., Hanaki, K., Suzuki, K., and Yamamoto, K. (2004) On the cyto-toxicity caused by quantum dots. *Microbiol. Immunol.* **48**, 669–675.

18. Larson, D. R., Zipfel, W. R., Williams, R. M., et al. (2003) Water-soluble quantum dots for multiphoton fluorescence imaging in vivo. *Science* **300**, 1434–1436.

19. Dickinson, M. E., Bearman, G., Tilie, S., Lansford, R., and Fraser, S. E. (2001) Multi-spectral imaging and linear unmixing add a whole new dimension to laser scanning fluorescence microscopy. *Biotechniques* **31**, 1272–1278.

20. Jaiswal, J. K., Goldman, E. R., Mattoussi, H., and Simon, S. M. (2004) Imaging with quantum dots. In: *Imaging in Neuroscience and development: A Laboratory Manual,* (Yuste, R. and Konnerth, A., eds.), Cold Spring Harbor Laboratory Press, New York, pp. 511–516.

9

Peptide-Mediated Intracellular Delivery of Quantum Dots

B. Christoffer Lagerholm

Summary

Quantum dots (QDs) have received a great amount of interest for use as fluorescent labels in biological applications. QDs are brightly fluorescent and very photostable, satisfying even imaging applications that require single molecule detection at high repetition rates over long periods of time (minutes to hours). There are by now numerous methods for conferring biospecificity and function including cell membrane permeability to QDs. A particular convenient method of conferring membrane penetrating ability and in some cases also biospecificity has been to couple biotinylated protein transduction domains to streptavidin-conjugated QDs. This method, which is easily customizable and requires minimal custom conjugation, is suitable for long-term in vitro and in vivo cell tracking imaging applications.

Key Words: Protein transduction domains; tat; quantum dots; nanocrystals; fluorescence imaging; streptavidin; biotin.

1. Introduction

The discovery of protein transduction domains (PTDs) has garnered much interest as a passive mean for drug delivery across cellular membranes, as well as across the blood–brain barrier (*1*). To date, a variety of PTDs have been used to deliver biologically relevant cargoes including peptides, proteins, oligonucleotides, radioactive tracers, liposomes, and magnetic nanoparticles into cells (*1*).

We and others have shown that PTDs can also be used for intracellular delivery of quantum dots (QDs) into a variety of cell types (**Table 1**) by coupling PTDs to QDs via a streptavidin–biotin link (*2–4*), covalently (*5*) by electrostatic adsorption of PTDs containing free cysteine residues, or by adsorption to the synthetic PTD Pep-1 (*6*). QD cell labeling has also been achieved by

From: *Methods in Molecular Biology, vol. 374: Quantum Dots: Applications in Biology*
Edited by: M. P. Bruchez and C. Z. Hotz © Humana Press Inc., Totowa, NJ

105

Table 1
Protein Transduction Domains Used in Conjunction With Quantum Dots

PTD	Peptide sequence	References
SV 40 T antigen	(PPKKKRKV)$_2$	*4*
	CSSDDEATADSQHSTPPKKKRKV	*9*
Cytochrome c oxidase subunit VIII	MSVLTPLLLRGLTGSARRLPVPRAKIHC	*9*
	MSVLTPLLLRGLTGSARRLPVPRAKIHWLC	*5*
Poly-Arginine	RRRRRRRRR	*2*
	RRRRRRRRRRRKC	*5*
HIV Tat	RKKRRYRRR	*3*
Pep-1	KETWWETWWTEWSQPKKKRKV	*6*

nonpeptide-mediated means including by coating QDs with albumin *(7)*, antibodies to cell surface markers, or by using avidin-conjugated QDs along with nonspecific biotinylation of cell surface markers *(8)*.

Of the approaches used for QD intracellular delivery, coupling PTDs to QDs via streptavidin–biotin is most easily performed, as this approach requires no custom conjugation and relies on commercially available streptavidin-conjugated QDs and readily available custom-designed biotinylated PTDs. Here, we describe a protocol for labeling Swiss 3T3 mouse fibroblasts with streptavidin QDs coupled to biotinylated polyarginine (biotin-R_9) PTDs. The results of this work have been published *(2)*.

2. Materials

2.1. QDs

1. Streptavidin-conjugated QDs. (Quantum Dot Corporation, Hayward, CA) (*see* **Note 1**).
2. 10 mM Sodium borate, pH 8.2 containing 1% bovine serum albumin (BSA).

2.2. Biotinylated Protein Transduction Domain

1. Biotinylated R_9 peptide or other biotinylated PTD (*see* **Note 2**).

2.3. Cell Culture

1. Swiss 3T3 mouse fibroblasts (ATCC CCL-92, American Type Culture Collection, Manassas, VA).
2. Dulbecco's-modified Eagle's medium (DMEM; Invitrogen, Carlsbad, CA) supplemented with 10% calf serum (Invitrogen).
3. 1X Solution of trypsin-EDTA (0.25% trypsin with EDTA 4Na) (Invitrogen).
4. 100-mm Diameter tissue culture dishes.

2.4. Analysis by Flow Cytometry

1. Flow cytometer (*see* **Note 3**).
2. Fluorescence filter sets (*see* **Note 4**).
3. Ham's F-12 medium with 15 m*M* HEPES and without phenol red (Invitrogen) supplemented with 3% BSA (see **Note 5**).

2.5. Analysis by Fluorescence Microscopy

1. Light microscope (*see* **Note 6**).
2. Fluorescence filter sets (*see* **Note 4**).
3. Glass-bottom tissue culture dishes (MatTek Corporation, Ashland, MA).
4. Ham's F-12 medium with 15 m*M* HEPES and without phenol red (Invitrogen) supplemented with 10% calf serum (*see* **Note 5**).
5. Dulbecco's phosphate buffered saline containing calcium and magnesium (D-PBS; Sigma, St. Louis, MO).
6. Solution of 40% paraformaldehyde (Electron Microscopy Sciences, Fort Washington, PA).

3. Methods

The described method for peptide-mediated intracellular delivery of QDs is compatible with both nonadherent and adherent cell types. The resulting cell-to-cell labeling uniformity is best evaluated by flow cytometry, whereas intracellular localization is easily evaluated by fluorescence microscopy of cells plated on glass-bottom culture dishes or glass cover slips. Although the protocol as described was designed for labeling Swiss 3T3 fibroblasts, we have also used it for labeling human osteoblast-like cells (MG63), human endothelial cells (HeLa), and primary chicken embryonic fibroblasts with similar results *(2)*.

Intracellular labeling in our specific case did not lead to nuclear localization but rather to QD localization to intracellular vesicles predominantly found in the perinuclear region *(2)*. Other researchers, using different QDs but in some cases similar PTDs, have seen at least partial nuclear localization *(4,5,9)* as well as mitochondrial localization when using a PTD with a mitochondrial targeting sequence *(5,9)*. The reason for this discrepancy in QD localization is not known but is likely a consequence of slight differences in QD coatings.

3.1. QD Labeling of Cells in Suspension

1. Grow Swiss 3T3 mouse fibroblasts in 100-mm tissue culture dish to 70% confluence. Prepare cell supension by aspirating DMEM media from cells. Rinse cells in 2 mL of trypsin solution for 15 s. Aspirate trypsin solution and add 1 mL fresh trypsin solution and incubate at 37°C until cells detach. Add 5 mL of DMEM media containing 10% calf serum and pellet cells by centrifugation at 50*g*. Resuspend cells in 5 mL of DMEM media containing 10% calf serum. Perform cell count to determine cell density. Keep cells on ice until use.

2. Combine 50 μL of 2 μM streptavidin-conjugated QDs with a 50-fold molar excess of R_9-biotin. Dilute to 100 μL in 10 mM sodium borate buffer, pH 8.2 containing 1% BSA. React for 10 min (*see* **Note 7**). The QDs used can either be of a single color or combinations of spectrally separated colors for multiplex labeling (*see* **Note 8**).
3. Prepare QD-labeling solution by mixing 2×10^5 cells with at least 25 pmol of preincubated streptavidin QD R_9-biotin complex and diluting to a final volume of 0.5 mL in DMEM media containing 10% calf serum.
4. Incubate at 4°C for 30 min while mixing.
5. Pellet cells by centrifugation at 50g and aspirate QD labeling solution.
6. Resuspend cells in DMEM media containing 10% serum and incubate cell suspension at 37°C for 1 h to allow for QD internalization.

3.2. QD Labeling of Adherent Cells

1. Seed 5×10^4 Swiss 3T3 mouse fibroblasts in a glass-bottomed tissue culture dish.
2. Combine 50 μL of 2 μM streptavidin-conjugated QDs with a 50-fold molar excess of R_9-biotin. Dilute to 100 μL in 10 mM sodium borate buffer, pH 8.2 containing 1% BSA. React for 10 min (*see* **Note 7**). The QDs used can either be of a single color or combinations of spectrally separated colors for multiplex labeling (*see* **Note 8**).
3. Prepare QD-labeling solution by diluting at least 10 pmoles of preincubated streptavidin QD R_9-biotin complex to 1 mL with DMEM containing 10% calf serum. Incubate at 4°C while mixing for 30 min (*see* **Note 9**).
4. Replace QD-labeling solution with DMEM containing 10% calf serum.
5. Incubate at 37°C for 1 h to allow for QD internalization.

3.3. Analysis by Flow Cytometry

1. For QD labeling of suspended cells: pellet cells by centrifugation at 50g and aspirate media. Resuspend cells in trypsin solution and incubate for 5 min at 37°C. Pellet cells by centrifugation at 500 rpm and aspirate trypsin solution. Resuspend cells in Ham's F-12 media with 3 mg/mL BSA and keep on ice until use.
2. For QD labeling of adherent cells: prepare cell supension by aspirating media. Rinse cells in 2 mL of trypsin solution for 15 s. Aspirate trypsin solution and add 1 mL fresh trypsin solution and incubate at 37°C until cells detach. Pellet cells by centrifugation at 50g and aspirate trypsin solution. Resuspend cells in Ham's F-12 media with 3 mg/mL BSA and keep on ice until use.
3. Install appropriate fluorescence filters (*see* **Note 4**).
4. Adjust high voltage and gain on each photomultiplier tube so as to maximize the signal above background in each channel while minimizing cross-talk between channels.
5. Analyze cell-to-cell uniformity of labeled cells by determining the mean fluorescence intensity and the standard deviation for each labeling condition.

3.4. Analysis by Fluorescence Microscopy

1. For QD labeling of suspended cells: seed cells in glass-bottom tissue culture dishes and allow the cells to attach and spread at 37°C.

2. For QD labeling of adherent cells: leave at 37°C until ready to proceed.
3. Image live cells in Ham's F-12 media containing 10% calf serum, or alternatively fix cells by washing once in D-PBS, incubate with 4% paraformaldehyde in D-PBS for 10 min, wash three times in D-PBS, and image in D-PBS (*see* **Note 10**).
4. Adjust exposure times in each color channel so as to maximize the signal above background while minimizing cross-talk between color channels.
5. Acquire images to determine QD intracellular location.
6. To display images, adjust the brightness and contrast values simultaneously on all images within a given color channel such that the fluorescence signal overlap of each color QDs into adjacent detector channels is minimized (**Fig. 1**) (*see* **Note 11**).

4. Notes

1. We have used QDs from Quantum Dot Corp. with peak emission wavelengths of 565, 605, and 655 nm. These QDs had streptavidin covalently conjugated to a water stabilizing QD coating composed of an amphiphilic co-polymer of octylamine and polyacrylic acid *(10)*. Streptavidin-conjugated QDs are also available from Antibodies Incorporated (Davis, CA). Alternatively, QDs available commercially from Evident Technologies (Troy, NY) or BioPixels (Westerville, OH) can be coupled to streptavidin.
2. Custom-synthesized biotinylated peptides are readily available from a large number of vendors. In the method described, Boc chemistry protocols using solid-phase techniques and commercially available Boc-N-α-t-Boc-N-ω-tosyl-D-arginine on a Rink amide MBHA resin were used to synthesize polyarginine–biotin conjugates with or without an additional 6-aminocaproic acid spacer (R_9-biotin; R_9-C_6-biotin) *(2)*.
3. We used an Epics Elite flow cytometer equipped with an argon laser tuned to the 488-nm line for fluorescence excitation and fluorescence bandpass emission filters centered at 565 (20-nm bandwidth), 610 (20-nm bandwidth), and 665 (45-nm bandwidth) nm, respectively (Chroma Technologies, Brattleboro, VT). The selected bandpass filters minimize, but do not completely eliminate, fluorescence signal overlap between the various QDs.
4. A significant advantage of QDs is the wide and overlapping absorption spectra of QDs with distinct emission spectras (for reviews on QD properties, *see* **refs.** *11,12*). As a result, QDs with distinct emission spectra may be simultaneously excited at wavelengths that are blue-shifted relative to the QD color with the lowest peak emission wavelength. In general, the QD fluorescence emission is maximized when using UV excitation, however, for live cells it is advisable to use visible excitation in the blue or green part of the spectra. We have successfully used an argon laser tuned to 488 nm for fluorescence excitation for flow cytometry and an excitation filter centered at 440 nm (20-nm bandwidth) for fluorescence microscopy. In addition, one should select emission filters with narrow bandwidths. We used QDs whose emission wavelength were centered at 565, 605, and 655 nm and fluorescence emission filters centered at 565 (20-nm bandwidth), 610 (20-nm bandwidth), and 654 (24-nm bandwidth) nm, respectively (Chroma Technologies).

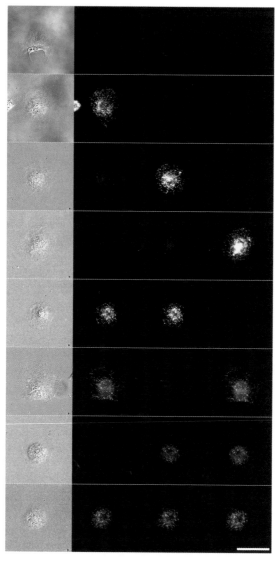

Fig. 1. Qualitative analysis of cell multicolor coding by fluorescence microscopy. Swiss 3T3 fibroblasts were labeled in suspension as described. Images shown are (column 1) differential interference contrast, (column 2) 565-nm fluorescence, (column 3) 605-nm fluorescence, and (column 4) 655-nm fluorescence. Cells were labeled with (row 1) blank control, (row 2) 565 quantum dots (QDs), (row 3) 605 QDs, (row 4) 655 QDs, (row 5) 565 and 605 QDs, (row 6) 565 and 655 QDs, (row 7) 605 and 655 QDs, (row 8) 565, 605, and 655 QDs. Fluorescence images within each column were scaled to the same brightness and contrast values (scale bar = 20 μm). (Reproduced with permission from **ref. 2** Copyright 2004 American Chemical Society.)

5. The pH indicator phenol red increases background fluorescence and should be avoided during fluorescence imaging and cytometry analysis.

6. We used a motorized Axiovert 135TV microscope (Zeiss, Thornwood, NY) equipped with a mercury arc lamp, a fluorescence excitation filter wheel, and a Hamamatsu ORCA-II charge-coupled device camera (Hamamatsu Photonics, Hamamatsu, Japan). The fluorescence excitation filter was placed in the excitation filter wheel while dichroic mirrors and emission filters were placed in adjacent positions in a filter slider. An improved microscope configuration (only requiring a single dichroic mirror) would have a fluorescence emission filter wheel for the emission filters and with the fluorescence excitation filter and accompanying dichroic mirror placed in a fluorescence filter slider or filter cube.

7. We always prepared the QD R_9–biotin complex fresh as significant QD aggregation occurred following the conjugation. We observed qualitatively less aggregation when using the R_9-biotin peptide as compared to the R_9-C_6-biotin peptide indicating that the increased hydrophobicity of the QD R_9-C_6-biotin complex may be at least partly responsible.

8. We have labeled Swiss 3T3 fibroblasts with all possible combinations of streptavidin QDs with peak emission wavelengths of 565, 605, and 655 nm and with a binary intensity scheme (two intensities, three colors for a total of 2^3 combinations). To compensate for differences in QD brightness, we used a 4:1:2 molar ratio of 565, 605, and 655 nm QDs, respectively.

9. QD labeling of adherent cells will result in a significant amount of nonspecific QD binding to the glass. To avoid this problem, we prefer to label cells in suspension as described and then allow the cells to attach and spread prior to fluorescence imaging analysis. As an alternative, it has been reported that the nonspecific QD binding to glass can be minimized by precoating the glass with polylysine (5).

10. We observed no differences in QD localization of live and fixed cells.

11. We used ImageJ, which is available as a free download at http://rsb.info.nih.gov/ij/, for image analysis and display. To display images for multiplex experiments, we opened all images for a particular color channel simultaneously and converted them to an image stack. We then adjusted the brightness and contrast values for the stack, such that fluorescence signal overlap of each color QDs into adjacent detector channels was minimized.

Acknowledgments

This original work was done in collaboration with the laboratories of A. S. Waggoner and D. H. Ly at Carnegie Mellon University and M. P. Bruchez and H. Liu at Quantum Dot Corporation and with support from NIH grant number R01 EB-000364. This chapter was prepared with support from NIH grant number R01 GM-41402.

References

1. Dietz, G. P. and Bähr, M. (2004) Delivery of bioactive molecules into the cell: the Trojan horse approach. *Mol. Cell Neurosci.* **27**, 85–131.

2. Lagerholm, B. C., Wang, M. M., Ernst, L. A., et al. (2004) Multicolor coding of cells with cationic peptide coated quantum dots. *Nano Letters* **4,** 2019–2022.
3. Gao, X., Cui, Y., Levenson, R. M., Chung, L. W., and Nie, S. (2004) In vivo cancer targeting and imaging with semiconductor quantum dots. *Nat. Biotechnol.* **22,** 969–976.
4. Chen, F. Q. and Gerion, D. (2004) Fluorescent CdSe/ZnS nanocrystal-peptide conjugates for long-term, non toxic imaging and nuclear targeting in living cells. *Nano Letters* **4,** 1827–1832.
5. Hoshino, A., Fujioka, K., Oku, T., et al. (2004) Quantum dots targeted to the assigned organelle in living cells. *Microbiol. Immunol.* **48,** 985–994.
6. Mattheakis, L. C., Dias, J. M., Choi, Y. J., et al. (2004) Optical coding of mammalian cells using semiconductor quantum dots. *Anal. Biochem.* **327,** 200–208.
7. Hanaki, K., Momo, A., Oku, T., et al. (2003) Semiconductor quantum dot/albumin complex is a long-life and highly photostable endosome marker. *Biochem. Biophys. Res. Commun.* **302,** 496–501.
8. Jaiswal, J. K., Mattoussi, H., Mauro, J. M., and Simon, S. M. (2003) Long-term multiple color imaging of live cells using quantum dot bioconjugates. *Nat. Biotechnol.* **21,** 47–51.
9. Derfus, A. M., Chan, W. C. W., and Bhatia, S. N. (2004) Intracellular delivery of quantum dots for live cell labeling and organelle tracking. *Adv. Mat.* **16,** 961–966.
10. Wu, X., Liu, H., Liu, J., et al. (2003) Immunofluorescent labeling of cancer marker Her2 and other cellular targets with semiconductor quantum dots. *Nat. Biotechnol.* **21,** 41–46.
11. Jaiswal, J. K. and Simon, S. M. (2004) Potentials and pitfalls of fluorescent quantum dots for biological imaging. *Trends Cell Biol.* **14,** 497–504.
12. Gao, X., Yang, L., Petros, J. A., Marshall, F. F., Simons, J. W., and Nie, S. (2005) In vivo molecular and cellular imaging with quantum dots. *Curr. Opin. Biotechnol.* **16,** 63–72.

10

Multiple Cell Lines Using Quantum Dots

Paul G. Wylie

Summary

The development of multiplexing capabilities and high-content readouts reporting individual cellular measurements enables assessment of biological variability on a single-cell basis, together with the evaluation of cell subpopulations within wells. A high-content screening multiplexed assay format allows additional information to be gained from a single assay. One such example is the ability to determine the effects of new chemical entities on different cell lines, tested in the same well. These assays, coupled with an appropriate automated cell-analysis platform, enable scalable screening of compound libraries for selectivity or toxicity. This approach can greatly increase screening efficiencies and enhance the amount of information achieved from a particular assay procedure, resulting in a significant reduction in the overall cost of a chemical compound library screen. By labeling live cells with Qtracker™ cell labeling kits and identifying cell proliferation using an Acumen Explorer® microplate cytometer, we were able to determine the differential rates of cell proliferation of the individual cell lines in the same well over time. This method can extend to multiplexing more than two cell populations and to measure drug-induced differential changes in proliferation in a single-well assay on multiple cell lines.

Key Words: Quantum dots; Qtracker; Acumen Explorer; high-content analysis; high-throughput screening; multiplexing; proliferation; fluorescence microplate cytometry.

1. Introduction

Determining differential responses of normal cells compared with cancerous cells in response to a single stimulus could prove valuable in selecting compounds that have maximal activity, with minimal toxicity to normal tissues. Such an assay could provide initial safety information at the level of a primary screen, removing acutely toxic compounds from consideration earlier in the screening process. The additional information that can be obtained using

From: *Methods in Molecular Biology, vol. 374: Quantum Dots: Applications in Biology*
Edited by: M. P. Bruchez and C. Z. Hotz © Humana Press Inc., Totowa, NJ

multiple cellular responses in an early screening process can be used to make more informed decisions, reducing the number of late dropouts in the drug development process. The transition from obtaining readouts from single-cell populations to multiple-cell populations in the screening process can transform the information content of these assays, just as the transition from biochemical-to-cellular assays has provided dramatic increases in the relevance of initial hits. For example, a utility of this methodology is to specifically identify the effects of compounds on cellular processes such as proliferation, cell death, apoptosis induction, kinase activation, and so on in multiple cell lines all in the same well. This approach could equally be extended to distinctly labeling a panel of knockout clonal cell lines and treating them in a single well with a single compound. There are many advantages of being able to gain this high level of information from more than one cell line per well. One such advantage is in high-throughput screening laboratories where cost is a fundamental concern. For example, the effect of a compound on a series of cells expressing different receptor subtypes can be analyzed in the same well. This approach can dramatically reduce the cost of a screen, as many cells types can be simultaneously tested in a single well. It is also able to reduce the quantity of valuable test compound used for each screen.

Quantum dots (QDs) absorb light over a broad spectral range and fluoresce at different wavelengths, which are determined by their physical size *(1,2)*. Qtracker™ cell labeling kits deliver fluorescent Qdot® nanocrystals (Quantum Dot Corporation, Hayward, CA) into the cytoplasm of live cells using a custom targeting peptide *(3,4)*. A specific enzyme does not mediate cytoplasmic delivery by this mechanism, therefore, cell-type specificity has not been observed. Once inside the cells, Qtracker labels provide intense, stable fluorescence that can be traced through several generations and are not transferred to adjacent cells in a population. Labeled cells can be observed for many days, with no cytotoxic, photobleaching, or degradation problems commonly associated with fluorescent dyes *(5,6)*.

In the method described here, we have used semiconductor QDs to achieve in-well labeling of distinct cell populations with Qtracker reagents. To analyze the subsequent growth of these cell populations, we used an Acumen Explorer (TTP LabTech Ltd, Melbourn, Hertfordshire, UK), a laser-scanning fluorescence microplate cytometer, which can simultaneously collect up to four colors of data in a single scan from cells in 96,384 and 1536 well plates. By combining the ability of Qtracker cell labeling reagents to be traced through several generations of cells, and the ability of an Acumen Explorer to rapidly scan live cells on a plate without affecting sterility, it is possible to analyze the same wells at each point throughout a time-course. This enables obtaining proliferation data from the same wells at each time-point. Because the same wells can be analyzed, the data quality is improved, as the inconsistency in the initial plating out of the cells is not an added variable. This simultaneous multiplexed detection of cell number can be run at plate read times as short as 5 min per 384-well plate and, therefore, provides the key requirements to enable a high-content assay to be used in a primary screening program.

2. Materials
2.1. Cell Culture

1. Dulbecco's modified Eagle's medium (Gibco/BRL) supplemented with 10% fetal calf serum (Gibco/BRL).
2. Minimum essential medium alpha (MEM-α) supplemented with 10% fetal calf serum (Gibco/BRL).
3. 5 mL 10,000 U/mL stock of penicillin is added to 500 mL culture medium to give a final concentration of 100 U/mL (Gibco/BRL).
4. 5 mL 10,000 μg/mL stock of streptomycin is added to 500 mL culture medium to give a final concentration of 100 μg/mL (Gibco/BRL).
5. 5 mL 200 mM stock of L-glutamine is added to 500 mL culture medium to give a final concentration of 2 mM (Gibco/BRL).
6. 25% Solution of trypsin and EDTA (0.38 g/L) (Gibco/BRL).

2.2. Cell Dye Loading

1. Tissue culture 24-well plate (BD Falcon).
2. Qtracker cell labeling kit 655- and 705-nm (Quantum Dot Corp.).
3. 96-Well assay plates, black walled/clear bottom (BD Falcon) (*see* **Note 1**).

3. Methods

Cell cultures grow at different rates according to the type and constitution of medium in which they are cultured. In the methods described here, two commonly used cells lines (CHO cells and SH-SY5Y cells) were grown independently in their respective standard culture medium, and then grown together in the same well in either standard CHO or SH-SY5Y medium. Prior to multiplexing the proliferation assay in the two cell lines, the compatibility of growing CHO cells and SH-SY5Y cells in different media was investigated. If alternative cell cultures are used to those described here, it is important to predetermine the effects of different media on those individual cell cultures.

Although Qdots fluoresce with very narrow spectral emissions and therefore there is only limited spectral overlap, it is still important to set up the assay using control wells that contain cells labeled with a single Qtracker color. The lack of spectral overlap of the Qtracker reagents makes them very easy to discriminate from each other and they are therefore ideal reagents to individually label multiple cell populations in the same well. In this method, two distinct cell lines have been used, but the same principle can be applied to more than just two. Color ratios can then be applied to identify the individual cell populations. Therefore, labeling of different cell lines with differing ratios of more than one Qdot conjugate enables the possibility of multiplexing an assay with many different cell lines together in a single well.

3.1. Subculture of Cells

1. CHO cells were routinely passaged (1:10–1:40 dilution) in MEM-α medium, supplemented by 10% (v/v) fetal calf serum, 100 U/mL penicillin, 100 µg/mL streptomycin, and 2 mM L-glutamine. Cells were passaged prior to achieving full confluency.
2. SH-SY5Y cells were routinely passaged (1:4) in Dulbecco's modified Eagle's medium and supplemented by 10% (v/v) fetal calf serum, 100 U/mL penicillin, 100 µg/mL streptomycin, and 2 mM L-glutamine. Cell lines were passaged prior to achieving full confluency (*see* **Note 2**).

3.2. Preparation of Cells Prior to Staining

1. The following protocol provides sufficient numbers of cells to run the final proliferation assay in all wells of a 96-well plate.
2. These instructions assume the use of a hemocytometer for accurately counting cell numbers.
3. Cells that had attained 80% confluency in a T75 tissue culture flask were carefully washed twice in prewarmed phosphate-buffered saline (PBS) and lifted by adding 5 mL trypsin/EDTA solution and incubating for 5 min at 37°C. The cell suspension was removed and spun down in a centrifuge for 10 min at 500g. The cells were resuspended in 5 mL of prewarmed cell culture medium and the number of cells determined by counting using a hemocytometer. The volume of the cell suspension was adjusted to achieve a final solution of 40,000 cells/mL with prewarmed cell culture medium.
4. Five hundred microliters of cell suspension was then aliquoted into each well of a 24-well tissue culture plate, achieving a final cell number of 20,000 cells/well (*see* **Note 3**).
5. The cells were allowed to adhere overnight onto the plate at 37°C in 5% CO_2.
6. Other cell lines can be used in a similar same manner (*see* **Note 4**).

3.3. Cell Labeling Solution

In this experiment, CHO cells were labeled with 655-nm Qtracker reagent and SH-SY5Y cells were labeled with 705-nm Qtracker reagent using the following method.

1. Prepare a 10 nM stock of labeling solution for each well of the 24-well plate to be labeled (*see* **Note 5**). Premix 1 µL of reagent A with 1 µL of reagent B (supplied in Qtracker cell labeling kit) in a 1.5-mL microcentrifuge tube (*see* **Note 6**).
2. Add 0.2 mL of fresh growth medium to the 1.5-mL tube and vortex for 30 s.

3.4. Cell Staining

1. Carefully aspirate the medium from the cells grown overnight in the 24-well plate.
2. Carefully add 0.2 mL of labeling solution to each well on the 24-well plate with cells. Take care not to disturb the cells from the base of the well.
3. Incubate the cells overnight at 37°C in 5% CO_2 (*see* **Note 7**).
4. The cells were washed twice with 500 µL fresh, prewarmed PBS. After washing, the cells were carefully lifted from the 24-well plate by adding 300 µL of trypsin/EDTA solution and incubated for 5 min at 37°C. The resultant cell suspension was removed from the well and spun down in a centrifuge for 10 min at 500g (*see* **Note 8**).
5. The number of cells was determined using a hemocytometer and the cells were resuspended in their respective prewarmed cell culture medium at a final concentration of 5000 cells/mL (*see* **Note 9**).
6. For this assay, CHO cells and SH-SY5Y cells were plated out individually into single wells to allow for accurate classification of each cell line (*see* **Note 10**). Therefore, 100 µL of each single cell solution was added to a well of a 96-well plate giving a final total of 500 cells/well. For the mixed cell population, 50 µL of each cell solution was added to each well of a 96-well microtitre plate giving 250 cells/well of CHO cells, and 250 cells/well of SH-SY5Y cells (this gives a final total of 500 cells/well). For a well view of the single and mixed cell populations (*see* **Fig. 1**).
7. The cells were incubated at 37°C and allowed to grow in the different media.

3.5. Scanning Cells

These instructions assume the use of an Acumen Explorer for cell analysis.

1. The cells were scanned using an Acumen Explorer fitted with filters from the QD range consisting of a 680 dichroic, a 655/20 bandpass filter (for CHO cells), and a 700/30 bandpass filter/mirror (for SH-SY5Y cells) (*see* **Note 11**). Scans were run on subsequent days to determine the respective cell numbers.
2. The scan settings of the Acumen Explorer were optimized to best detect both cell lines and increase the speed of analysis (*see* **Note 12**).

3.6. Cell Analysis

1. Every fluorescent object in the well was identified and cells were categorized from other fluorescent objects, such as debris, using size classifiers. Both cell lines were classified as having a width and depth of greater than 8 µm, as typical debris is significantly smaller than this value.
2. By simultaneously scanning for both Qtracker reagents, the ratio of 655 to 705-nm fluorescence was quantified and, therefore, the labeled CHO cells were easily discriminated from labeled SH-SY5Y cells in the same well (**Fig. 2**).
3. By using the different Qtracker labels to identify individual cells, the exact number of each cell type in each well was determined each day. An example of the results that was produced is shown in **Figs. 3** and **4**.

Fig. 1. Well views of single cell populations and mixed cell populations. (**A**) CHO cells labeled with a Qtracker 655 Kit. (**B**) SH-SY5Y cells labeled with a Qtracker 705 Kit. (**C**) CHO cells and SH-SY5Y cells together in a mixed population. Cells are pseudo-colored for clarity.

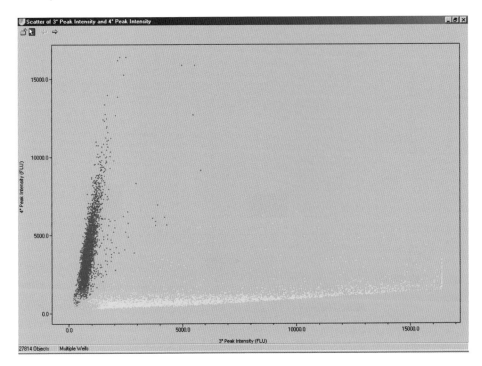

Fig. 2. Quantification of CHO cells and SH-SY5Y cells in a mixed cell population. By simultaneously scanning for both Qtracker reagents, the ratio of 655 to 705-nm fluorescence can be quantified and CHO cells (yellow) can be easily discriminated from SH-SY5Y cells (red) in the same well.

4. Notes

1. There is a certain amount of laser light that is reflected back into the light path, which adds to the background fluorescence detected. The use of black walled microplates is recommended as they reduce the amount of reflected laser light and improve the quality of the data obtained. Also, as the scanning laser within the Acumen Explorer is directed at the base of the microplate, flat clear-bottomed plates must be used. Although the method described here is for a 96-well plate, this method is equally applicable for use in both 384- and 1536-well microplates.

2. When culturing SH-SY5Y cells, the cells should never be allowed to attain full confluency, as this predisposes them to differentiating into neuronal cells. We have found, however, that the cells should not be seeded more than at a 1:4 dilution, as this can result in the SH-SY5Y cells taking a long time to recover and grow at the standard rate.

3. The loading of Qtracker reagent into cells is very dependent on cell numbers in the well. We have found that 20,000 cells/well in a 24-well plate is the optimal concentration for dye uptake, which also obtains a large number of cells for use in

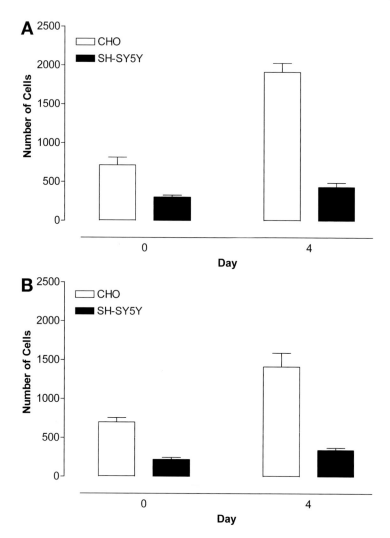

Fig. 3. Proliferation of CHO and SH-SY5Y cells in CHO media (MEM-α). (**A**) Cells grown in individual wells in CHO media in separate wells. (**B**) Cells grown in the same well in CHO media.

   an assay. When the cells were plated out at a higher density, such as at 100,000 cells/well, dye loading was less efficient.
4. This protocol can be used for many other cell culture systems. For cells that grow in suspension, such as Jurkat cells, the cells are treated as described, except that they are not lifted from the well by using trypsin/EDTA solution, but by simple aspiration from the well.
5. The dye concentration of Qtracker reagent should, for each cell line used, be fully optimized. The use of too concentrated a solution can result in poor dye uptake in

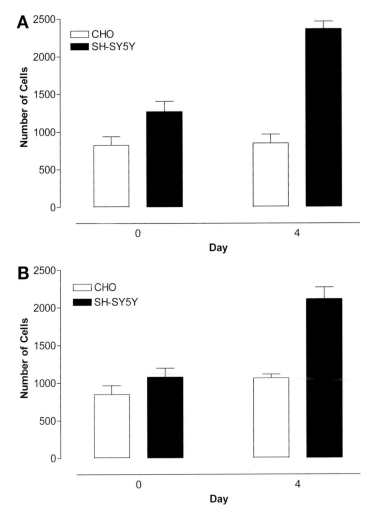

Fig. 4. Proliferation of CHO and SH-SY5Y cells in SH-SY5Y media (MEM-α). **(A)** Cells grown in individual wells in SH-SY5Y media in separate wells. **(B)** Cells grown in the same well in SH-SY5Y media.

certain cell lines, whereas a too dilute concentration can result in either poor uptake of the dye or increased incubation times.

6. To obtain enough cells to run the proliferation assay in every well of a 96-well plate, 3 wells of each cell line in a 24-well plate need to be labeled with Qtracker. To obtain optimum mixing of reagent A and B, when preparing the Qtracker label, it is important that the labeling solution should be prepared as described, i.e., prepare an individual sample for each well of the 24-well plate that is to be labeled (e.g., if three wells are to be labeled, three microcentrifuge tubes with a final volume of 0.2 mL are required).

7. In this method, we found that incubating cells for 1 h did not produce enough loading of the dye into the cells to allow proliferation studies to be run over several days, as the amount of dye decreases with each cell division. However, a 1-h incubation did stain the cells well enough to be easily observed under a fluorescence microscope. We routinely found that an overnight incubation gave the best results, although we also gained adequate staining using a 5-h incubation time to enable accurate enumeration on an Acumen Explorer.

8. Although SH-SY5Y cells do not attach to the plate very firmly, it is very easy to aspirate and lose a substantial number of the cells. To prevent this cell loss, it is necessary to retain all the PBS from wash steps and add to the trypsinised cells for spinning in the centrifuge.

9. At this stage, the cells can be very clumpy after resuspending them in medium following centrifugation. It is important to ensure that the cells are separated adequately to make certain of accurate counting of the cells in the proliferation study. This is easily achieved by using a 5-mL syringe attached to a 20 gage, 1.5-in. polypropylene needle; carefully pipet up and down three times, minimizing the amount of foaming. Increased foaming of the sample can cause shearing forces to disrupt cell membranes and therefore results in cell loss.

10. By having control wells in which cells labeled with a single color have been aliquoted, the initial identification and subsequent classification of individual cell populations is made much simpler. Once the parameters have been established and the labeling characteristics become well known, this step is no longer required.

11. We found that using the custom-built filter sets described were more efficient at discriminating between the labeled cells compared with the standard Acumen Explorer filter sets.

12. When setting up the assay, initial scans were run using a resolution value of 1 by 4 μm, with a scanned area of 49 mm^2/well. These settings gave a plate read time of 15 min. Faster plate read times were obtained by scanning with a resolution of 1 by 8 μm, and a scanned area of 16 mm^2/well, giving read and analysis times in the order of 5 min/plate.

Acknowledgments

The author would like to thank Dr. Marcel Bruchez at Quantum Dot Corporation for supplying the Qtracker reagents, and Dr. John Budd, formerly of TTP LabTech Ltd, for his advice and technical assistance in the preparation of this text.

References

1. Aliviastos, A. P. (1996). Semiconductor clusters, nanocrystals and quantum dots. *Science* **271**, 933–937.
2. Bruchez, M. P., Moronne, M., Gin, P., Weiss, S., and Aliviasatos, A. P. (1998) Semiconductor nanocrystals as fluorescent biological labels. *Science*. **281**, 2013–2016.
3. Jaiswal, J. K., Mattoussi, H., Mauro, J. M., and Simon, S. M. (2003) Long-term multiple color imaging of live cells using quantum dot bioconjugates. *Nature Biotech*. **21**, 47–51.

4. Hanaki, K., Momo, A., Oku, T., et al. (2003) Semiconductor quantum dot/albumin complex is a long-life and highly photostable endosome marker. *Biochem. Biophys. Res. Comm.* **302,** 496–501.
5. Futaki, S., Goto, S., Suzuki, T., Nakase, I., and Sugiura, Y. (2003). Structural variety of membrane permeable peptides. *Curr. Protein Pept. Sci.* **4,** 87–96.
6. Umezawa, N., Gelman, M. A., Haigis, M. C., Raines, R. T., and Gellman, S. H. (2001) Translocation of a beta-peptide across cell membranes. *J. Am. Chem. Soc.* **124,** 360–369.

11

Measuring Cell Motility Using Quantum Dot Probes

Weiwei Gu, Teresa Pellegrino, Wolfgang J. Parak, Rosanne Boudreau, Mark A. Le Gros, A. Paul Alivisatos, and Carolyn A. Larabell

Summary

The ability of cancer cells to migrate and metastasize is known to be directly related to tumor cell motility. Therefore, assaying the level of tumor cell motility is an excellent indicator of metastatic potential. We have developed an efficient and sensitive two-dimensional cell motility assay to image the phagokinetic uptake of colloidal CdSe/ZnS semiconductor nanocrystals (quantum dots [QDs]).

As cells move across a thin, homogeneous layer of QDs, they engulf and uptake the nanocrystals and leave behind a fluorescent-free trail. By measuring the ratio of trail area to cell area we have discovered that it is possible to distinguish between noninvasive and invasive cancer cells lines. This technique has, therefore, the potential to be used as a rapid, robust, and quantitative in vitro measure of metastatic potential. Because the technique only relies on fluorescence detection, requires no significant data processing, and is used with live cells, it is both rapid and straightforward.

Key Words: Fluorescent semiconductor nanocrystals; quantum dots; cell motility; cell migration; motility assay.

1. Introduction

The level of tumor cell motility and migration has been shown to correlate directly with the invasive potential of cancer cells (1). Several assays and techniques have been used to observe or quantify cell motility: the Boyden chamber assay (2–4), time lapse video (5), and phagokinetic tracking (6,7). However, these assays suffer from a number of practical limitations. For example, in the most widely used assay, the Boyden chamber assay, the passage of cells through a membrane is monitored. Consequently, in addition to being laborious, this assay does not allow for real-time variations in conditions, and in some cases the assay fails to adequately take account for the complex in vivo environment.

From: *Methods in Molecular Biology, vol. 374: Quantum Dots: Applications in Biology*
Edited by: M. P. Bruchez and C. Z. Hotz © Humana Press Inc., Totowa, NJ

The phagokinetic tracks method, proposed by Albrecht-Buehler, is a powerful technique that integrates cell motility measurements with the history of individual cell paths. In concept the method is beautifully simple—let cells record their own movements by clearing tracks as they pass over a layer of "markers."

We have developed a two-dimensional in vitro cell motility assay based on the phagokinetic uptake of colloidal fluorescent CdSe/ZnS nanocrystals (quantum dots [QDs]) *(8,9)*. Compared to the Boyden chamber invasion assay, the QD-based cell motility assay has greater sensitivity and can rapidly discriminate between noninvasive and invasive tumor cells. It uses fluo-rescence detection, requires no processing, and can be used in live cell studies.

Colloidal CdSe/ZnS semiconductor nanocrystals are inorganic, photochemically robust fluorophores *(10–13)*. When excited with ultraviolet (UV) light, fluo-rescence in the visible range is emitted. The emission color can be tuned by simply changing the size of the particle. When cancer cells are seeded on top of a thin homogenous layer of nanocrystals, they engulf the nanocrystals as they move and leave behind a phagokinetic track free of nanocrystals (**Fig. 1**). The relative area of the phagokinetic track with respect to the area of the cell is then analyzed. Cells with different metastatic potentials leave dramatically different fluorescence free tracks along their migration path. Invasive cancer cells leave a large fluorescence-free area in their wake, usually larger than the cell area, whereas noninvasive cancer cells only remove a small area of nanocrystals around their periphery. Statistics on seven cell lines have shown that the average ratio of (trail area)/(cell area) is greater than one for invasive and metastatic cell lines, whereas metastatic but in vitro noninvasive cells have an average ratio less than one.

QDs for this cell motility assay can either be synthesized in the lab or are commercially available. The variability of the assay between different QDs has not been assessed.

2. Materials

2.1. QD Layer Preparation

1. CdSe/ZnS silica-coated QDs with emission at 620 nm were synthesized as described previously *(14)* (other types of water-soluble QDs can be obtained from Quantum Dot Corp. (Hayward, CA) or Evident Technologies (Troy, NY).
2. Type I collagen 2.9 mg/mL collagen I in 12 µ*M* HCl (Cohesion, Palo Alto, CA).
3. Autoclaved or sterile-filtered phosphate-buffered saline: 137 m*M* NaCl, 2.7 m*M* KCl, 4.3 m*M* Na$_2$HPO$_4$, and 1.4 m*M* KH$_2$PO$_4$, pH 7.3.
4. Autoclaved distilled water.
5. Sterile four-well chambered cover glass slide (LabTEK II, Campbell, CA).

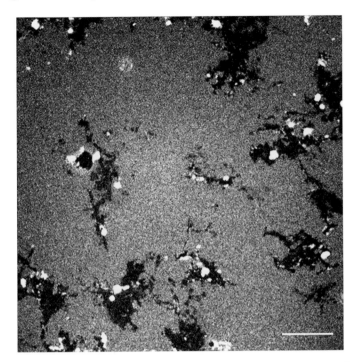

Fig. 1. Phagokinetic track of the highly metastatic human mammary gland adeno-carcinoma cell line MDA-MB-231 grown on a collagen layer that had been coated with a layer of silica-coated, water-soluble fluorescence quantum dots. While cells migrated across the layer, they engulfed the nanocrystals and left areas that were fluorescence-free. The scale bar is 200 μm. Image was collected with a confocal microscope using a fluorescence detector to record the nanocrystal trail. (Reproduced with minor modification from **ref.** *8* with permission from Wiley Publishing.)

2.2. Cell Culture

1. Cell lines (i.e., human breast cancer cell line MDA-MB-231).
2. Cell culture media (i.e., Leibovitz's L-15 [ATCC, Manassas, VA]) and any required additives (i.e. fetal bovine serum (Invitrogen, Carlsbad, CA).
3. 0.25% Trypsin solution.

2.3. Confocal Microscopy

1. Bio-Rad MCR-1024 (Bio-Rad) laser scanning confocal imaging system with Nikon Diaphot and a Fluor ×20, 0.75 NA lens or similar confocal imaging system. Or, Zeiss LSM 510 NLO microscope system configured with an inverted microscope using a ×40, 0.6 NA, long working distance water immersion lens custom-fitted with a water reservoir, or a ×40, 1.2 NA water immersion lens. Adjust the filter sets according to the QDs. Laser excitation wavelength of 488 nm and long-pass

emission filter at 585 nm can be used for dots with an emission wavelength at 620 nm.

3. Methods

This in vitro cell motility assay is based on observation and quantitation of the phagokinetic uptake of colloidal fluorescent QDs. The procedure is comprised of three steps. First, a thin homogenous QD layer is prepared, then the cells are plated on top of the layer, and after a period of incubation the cells are imaged and the motility tracks are analyzed.

3.1. Preparation of the QD Layer

1. Prepare collagen-coating buffer. Add type I collagen stock solution to phosphate-buffered saline (pH 7.3) for 1:44 dilution at 4°C.
2. Add collagen-coating buffer (1 mL per well) to the four-well chambered cover glass slide, then incubate chamber at 37°C for 1 h.
3. Carefully aspirate the residual collagen solution without disturbing the coating. The collagen-coated chamber slide can be used immediately, or stored at 4°C for use within 24 h.
4. Prepare QD working solution. Dilute silica shell-coated CdSe/ZnS stock solution (*see* **Note 1**) in distilled water to a final concentration of 0.2 µ*M*.
5. Add 15 µL working stock QDs solution to the collagen layer prepared in **step 3**, and dry under a UV lamp in a sterile hood for 30 min. The QDs layer should be used immediately after this step, and not stored (*see* **Notes 3 and 4**).

3.2. Cell Seeding on the QD Layer

1. Trypsinize cells at approx 80% confluence, and count the cell number.
2. The recommended plating density is 1000 cells/cm². Pellet the appropriate amount of cells by centrifugation at 400 rcf at room temperature for 5 min. Discard the supernatant.
3. Resuspend the cells with 2-mL tissue culture media. Add 0.5 mL/well on top of the QDs layer. Incubate the cells at 37°C for 24 h (*see* **Note 5**).

3.3. Imaging and Assay

1. After incubation, images were taken using a confocal microscope at a magnification of ×20. For QDs emitting at 620 nm, an excitation wavelength of 488 nm and a long-pass emission filter at 585 nm can be used. Tune the laser power to the lowest possible level that still allows imaging of the QD fluorescence. At least 10 images at different locations in the well are required to provide enough cell tracks so that the results are statistically significant. An example image is shown in **Fig. 1**.
2. One way of analyzing the cell motility is to analyze the cell track (fluorescence-free) area vs the area of the corresponding cell. To facilitate rapid data processing, Image Pro Plus software or similar image processing software can be used with automated script language procedures. Examples of the cell motility analyses are shown in **Fig. 2** (*see* **Note 6**).

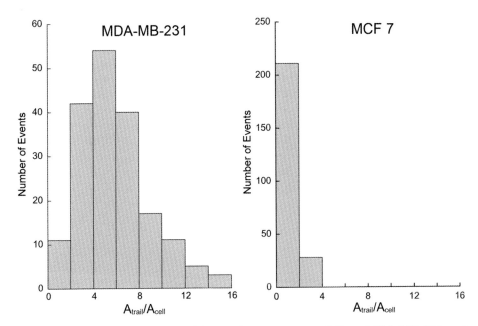

Fig. 2. Statistics on motility of human breast cancer cell lines MDA-MB-231 and MCF 7. The plots show the ratio (trail area)/(cell area) vs the frequency of events. The average values for the most probable size of the trail per cell line estimated are: MDA-MB-231, 6 ± 3 (SD) on 182 pictures and MCF 0.9 ± 0.7 (SD) on 239 pictures.

4. Notes

1. QDs are best stored at 4°C in the dark. The QDs working solution should be used as fresh as possible. Storing the working solution at these low concentrations might result in QD precipitation.
2. Filtering the QDs with a 0.2-μm filter should prevent bacterial contamination carried through from the synthesis procedure (which is normally carried out in non-sterile conditions). However, if this is not feasible, for example if the sample is found to stick to the filter membrane, or there is only a small volume of QDs required, the QDs layer can be sterilized by placing under a UV lamp in a sterile hood overnight. If possible, the addition of antibiotics (50 U/mL penicillin and 50 μg/mL strepto-mycin) to the tissue culture media also can minimize contamination.
3. Cracks on the QD layer are indicative that the QD layer was not used soon enough after preparation. To avoid this happening, use the layer as fresh as possible.
4. To ensure the QD layer is as homogeneous as possible, use a fresh collagen layer. When the collagen layer is ready, aspirate the residual collagen buffer but leave a small amount of solution on the layer surface, then immediately add the QD solution.
5. The use of phenol red-free media vs media containing phenol red can affect the motility behavior of some cells. We noticed that the human breast epithelial cell line

MCF 10A leaves a fluorescence-free trail on the QD layer when grown in media containing phenol red, but they do not leave trails when grown in phenol red-free media.

6. To obtain statistically significant results, a number of images for each cell line should be collected and then a histogram plot of the ratio of (trail area)/(cell area) vs the frequency of events drawn (**Fig. 2**). An average value for the most probable size of the trail per cell line can then be estimated. We compared seven cell lines using this assay. Under the conditions of this assay, for invasive and metastatic cell lines, the average ratio of track area to cell area was greater than one, whereas for metastatic noninvasive cells the same ratio is, on average, less than one.

7. To facilitate analysis, image processing can be automated with the use of script language in the Image Pro Plus program or similar software. The software can automatically outline a cell and the corresponding fluorescence-free cell track, and calculate the ratio of these areas.

Acknowledgments

This work was supported by NIH National Center for Research Resources through the Univ. of California, Los Angeles subaward agreement 0980GFD623 through the US Department of Energy under contract no. DE-AC03-76SF00098.

References

1. Partin, A. W., Schoeniger, J. S., Mohler, J. L., and Coffey, D. S. (1989) Fourier analysis of cell motility correlation of motility with metastatic potential. *Proc. Natl. Acad. Sci. USA* **86,** 1254–1258.
2. Albini, A., Iwamoto, Y., Kleinman, H. K., et al. (1987) A rapid *in vitro* assay for quantitating the invasive potential of tumor cells. *Cancer Res.* **47,** 3239–3245.
3. Kramer, R. H., Bensch, K. G., and Wong, J. (1986) Invasion of reconstituted basement membrane by metastatic human tumor cells. *Cancer Res.* **46,** 1980–1989.
4. Terranova, V. P., Hujanen, E. S., and Martin, G. R. (1986) Basement membrane and the invasive activity of metastatic tumor cells. *J. Nat. Cancer Inst.* **77,** 311–316.
5. Rajah, T. T., Abidi, S. M. A., Rambo, D. J., Dmytryk, J. J., and Pento, J. T. (1998) The motile behavior of human breast cancer cells characterized by time-lapse videomicroscopy. *In Vitro Cell. Dev. Biol. Anim.* **34,** 626–628.
6. Albrecht-Buehler, G. (1977) The phagokinetic tracks of 3T3 cells. *Cell* **11,** 395–404.
7. Albrecht-Buehler, G. (1977) The phagokinetic tracks of 3T3 cells: parallels between the orientation of track segments and of cellular structures which contain actin or tubulin. *Cell* **12,** 333–339.
8. Parak, W. J., Boudreau, R., Le Gros, M., et al. (2002) Cell motility and metastatic potential studies based on quantum dot imaging of phagokinetic tracks. *Adv. Mat.* **14,** 882–885.
9. Pellegrino, T., Parak, W. J., Boudreau, R., et al. (2003) Quantum dot-based cell motility assay. *Differentiation* **71,** 542–548.

10. Murray, C. B., Norris, D. J., and Bawendi, M. G. (1993) Synthesis and characterization of nearly monodisperse CdE(E=S, Se, Te) semiconductor nanocrystallites. *J. Amer. Chem. Soc.* **115,** 8706–8715.

11. Hines, M. A. and Cuyostsionnest, P. (1996) Long-term multiple color imaging of live cells using quantum dot bioconjugates. *Nat. Biotechnol.* **21,** 47–51.

12. Dabbousi, B. O., Rodriguez-Viejo, J., Mikulec, F. V., et al. (1997) (CdSe)ZnS core-shell quantum dots: synthesis and characterization of a size series of highly luminescent nanocrystallites. *J. Phys. Chem. B* **101,** 9463–9475.

13. Alivisatos, A. P. (1996) Semiconductor clusters, nanocrystals and quantum dots. *Science* **271,** 933–937.

14. Gerion, D., Pinaud, F., Williams, S. C., et al. (2001) Synthesis and properties of biocompatible water-soluble silica-coated CdSe/ZnS semiconductor quantum dots. *J. Phys. Chem. B* **105,** 8861–8871.

III

IMAGING OF LIVE ANIMALS

12

Quantum Dots for In Vivo Molecular and Cellular Imaging

Xiaohu Gao, Leland W. K. Chung, and Shuming Nie

Summary

Multifunctional nanoparticle probes based on semiconductor quantum dots (QDs) are developed for simultaneous targeting and imaging of cancer cells in living animals. The structural design involves encapsulating luminescent QDs with an ABC triblock copolymer, and linking this polymer to tumor-targeting ligands, such as antibodies and drug-delivery functionalities. In vivo targeting studies of human prostate cancer growing in nude mouse show that the QD probes can be delivered to tumor sites by both enhanced permeation and retention (passive targeting) and by antibody binding to cancer-specific cell surface biomarkers such as prostate-specific membrane antigen (active targeting). Using both subcutaneous injection of QD-tagged cancer cells and the systemic injection of multifunctional QD probes, multicolor fluorescence imaging of as few as 10–100 cancer cells can be achieved under in vivo conditions. The use of spectrally resolved imaging can efficiently remove autofluorescence background and precisely delineate weak spectral signatures in vivo. These results suggest that QD probes and spectral imaging can be combined for multiplexed imaging and detection of genes, proteins, and small-molecule drugs in single living cells, and that this imaging modality can be adopted for real time visualization of cancer cell metastasis in live animals.

Key Words: Quantum dots; nanoparticles; in vivo; molecular; cellular; imaging; targeting; diagnosis; spectral; multiplexed; multifunctional; block copolymer.

1. Introduction

The development of high-sensitivity and high-specificity probes beyond the intrinsic limitations of organic dyes and fluorescent proteins is of considerable interest to many areas of research, ranging from molecular and cellular biology to molecular imaging and medical diagnostics. Recent advances have shown that nanometer-sized semiconductor particles can be covalently linked with

From: *Methods in Molecular Biology, vol. 374: Quantum Dots: Applications in Biology*
Edited by: M. P. Bruchez and C. Z. Hotz © Humana Press Inc., Totowa, NJ

biorecognition molecules such as peptides, antibodies, nucleic acids, and small-molecule inhibitors for applications as fluorescent probes *(1–13)*. In comparison with organic fluorophores, these quantum-confined particles or quantum dots (QDs) exhibit unique optical and electronic properties, such as size- and composition-tunable fluorescence emission from visible-to-infrared wavelengths, extremely large absorption coefficients across a wide spectral range, and very high levels of brightness and photostability *(14,15)*. Despite their relatively large sizes (2–8 nm), recent research has shown that bioconjugated QD probes behave like genetically encoded fluorescent proteins (4–6 nm), and do not suffer from serious kinetic binding or steric-hindrance problems *(6–13)*. In this "mesoscopic" size range, QDs also have more surface areas and functionalities that can be used for linking to multiple diagnostic (e.g., radioisotopic or magnetic) and therapeutic (e.g., anticancer) agents. These properties have opened new possibilities for ultrasensitive bioassays and diagnostics, as well as for advanced molecular and cellular imaging.

Here, we report detailed protocols of preparing bioconjugated QD probes for simultaneous targeting and imaging of human prostate cancer cells in a murine model. Key steps involves high-quality QD preparation, surface coating with amphiphilic triblock copolymer for in vivo protection, bioconjugation of multiple polyethylene glycols (PEGs) and targeting ligands for tumor antigen recognition, and in vivo fluorescence imaging. To enhance the detection sensitivity, we further discuss the use of hyperspectral imaging configuration to separate QD fluorescence from strong background (mouse skin autofluorescence).

2. Materials

1. 90% Technical grade trioctylphosphine oxide (TOPO) (Aldrich, St. Louis, MO).
2. 99% Pure trioctylphosphine (Aldrich).
3. Cadmium oxide (CdO 99.99%) (Aldrich).
4. Selenium (>99%) (Riedel-de Haën, through Aldrich).
5. 99% Stearic acid (Sigma).
6. Hexamethlydisilathiane (Fluka, through Aldrich).
7. Dimethylzinc (10% wt in hexane, store and use in inert atmosphere) (Strem).
8. 98% Hexadecylamine (Aldrich).
9. Poly(t-butyl acrylate-co-ethyl acrylate-co-methacrylic acid) (PBEM, mw. 100K) (Aldrich).
10. 99% Octylamine (Fluka).
11. *N*-(3-dimethylaminopropyl)-*N*′-ethylcarbodiimide hydrochloride (EDAC) (Fluka).
12. PEG (mw. 2-5K) (Nektar, Huntsville, AL and Sunbio, Orinda, CA).
13. 98% 2,2′-(Ethylenedioxy)diethylamine (Aldrich).
14. Antibody J591 against prostate-specific membrane antigen (PSMA) (Millennium Pharmaceuticals, Cambridge, MA).
15. Separation media, Sephadex G-25, Superdex 75, and Superdex 200 (Amersham).

16. Ketamine and Xylazine (Prescription drugs from local hospital).
17. Ultracentrifuge, Optima TLX (Beckman Coulter, Fullerton, CA).
18. 1-cc Insulin syringe (29X1/2 gauge) for intravenous injection (VWR, West Chester, PA).
19. 6–8-wk nude mice (Charles River, Wilmington, MA).
20. Specialized imaging equipment is discussed in the main text.

3. Methods

3.1. Preparation of Highly Fluorescent QDs

High-quality red-color QDs are prepared according to literature procedures with modifications *(16–20)*. The 0.128 g CdO (1 mmol) precursor is first dissolved in 1 g stearic acid with heating in a three-neck round-bottom flask. After formation of a clear solution, TOPO (5 g) and hexadecylamine (5 g) are added as reaction solvents, which are then heated to 250°C under argon for 10 min. The temperature is briefly raised to 360°C, and equal molar selenium dissolved in trioctylphosphine is quickly injected into the hot solvents. The mixture immediately changes color to orange-red, indicating QD formation. The dots are kept in the reaction solvents at 200°C for 30 min, and capping solution of dimethylzinc (0.5 mmol) and hexamethyldisilathiane (0.5 mmol) is slowly added over a period of 15 min to protect the CdSe core. These ZnS-capped CdSe dots have excellent chemical- and photostability. The dots are cooled to room temperature, and are rinsed repeatedly with methanol/hexane to remove free ligands. Ultraviolet adsorption, fluorescence emission spectra, and transmission electron microscopy are used for characterization. This procedure typically produces QDs with emission peak centered at 630–640 nm, close to the upper wavelength limit of high-quality CdSe dots. However, the deep-red color is not optimized for tissue penetration and imaging sensitivity in animals. Deep tissue imaging (millimeters to centimeters) requires the use of near-infrared light in the spectral range of 700 to 900 nm *(21)*, (*see* **Note 1**). Nevertheless, the imaging concept and probe preparation techniques are essentially the same.

3.2. Nanoparticle Surface Modification and Bioconjugation

3.2.1. Polymer Modification

For encapsulating QDs, about 25% of the free carboxylic acid groups in PBEM triblock copolymer are derivatized with octylamine, a hydrophobic side chain (*see* **Note 2**). Thus, the original polymer (0.1 g) dissolved in 4 mL dimethylformamide is reacted with *n*-octylamine (5.2 mg) using ethyl-3-dimethyl amino propyl carbodiimide (EDAC, 23 mg, threefold excess of *n*-octylamine), as a crosslinking reagent. The product yields are generally greater than 90% because of the high EDAC-coupling efficiency in dimethylformamide (determined by a change of the free octylamine band in thin-layer chromatography). The reaction mixture is dried with a rotary evaporator (Rotavapor R-3000, Buchi Analytical Inc., DE). The resulting oily liquid is precipitated and rinsed five times with water to remove excess EDAC and other water-soluble byproducts. After vacuum-drying, the octylamine-grafted polymer is stored or resuspended in an ethanol/chloroform mixture for use.

3.2.2. Particle Encapsulation

TOPO-capped purified QDs (0.1 nmol) are mixed with the polymer in a chloro-form/ethanol solvent mixture (3:1 [v/v]). The nanoparticle suspension is then placed in vacuum and slowly dried over a time course of 2–6 h for particle–polymer self-assembly. The polymer-to-QD molar ratio is set at 5–20 depending on the particle sizes (for the red QDs used here, the ratio is set at 20), and the polymers in excess are removed later. After vacuum-drying, the encapsulated dots are soluble in many polar solvents, such as aqueous buffer (pH >9.0) and alcohols. The nanoparticle–polymer hybrids are kept in aqueous solution for 3 d and then purified from unbound polymers by gel filtration (Superdex 200) (sample loading volume <5% of the column volume). Alternatively, ultracen-trifuge and ultrafiltration work equally well. Dynamic light scattering measure-ments show a particle size around 10 nm (number weighted), which is much smaller than nanoparticles coated with the original unmodified PBEM polymer (40 nm). This comparison indicates the formation of a tight polymer wrapping layer on QD surface.

3.2.3. Long-Circulating PEG-Modified QD

Polymer-coated QDs are activated with 50 mM EDAC in phosphate-buffered saline (PBS) and reacted with amino-mPEG (mw 5000) at a QD/PEG molar ratio of 1:50 overnight at pH 8.5 (pH adjusted by NaOH). The QDs saturated with PEG chains can be purified by three methods, column filtration (Superdex 75), dialysis (molecular weight cut-off [mwco] >3X of the mw of PEG), or ultracentrifugation at 75,000g for 60 min. After resuspension in PBS buffer (pH 7.4), trace amount of aggregated particles were removed by centrifugation at 6000g for 10 min. The resulted QDs not only have a long plasma circulation time but are also highly stable in a broad range of aqueous conditions (e.g., pH 1.0 to 14.0 and salt concentration 0.01 to 1 M).

3.2.4. QD–Antibody Probe

We have developed two coupling procedures based on carbodiimide-mediated amide formation and amine-sulfhydryl crosslinking (*22*). For carbo-diimide reactions, the polymer-coated dots (COOH functional groups) are activated with 1 mM EDAC for 10 min and then mixed with amino-mPEG at a QD/PEG ratio of 1:6. After a quick purification with polyacrylamide-desalting columns (Pierce), the activated dots are reacted with an IgG antibody at a QD/antibody molar ratio of 1:15 for 2 h. The final QD bioconjugates are purified by filtration column chromatography (Amersham). After dilution in PBS buffer, aggregated particles are removed by centrifugation at 6000g for 10 min, and the QD–antibody bioconjugates are kept at 4°C. This procedure is easy to perform and broadly applicable for many native proteins, such as IgG, streptavidin, lectins, peptides, and so on because the availability of amine groups (**Note 3**). On the other hand, however, the abundant reactive groups could cause aggregation and render biomolecules randomly oriented on the QD surface, which is detrimental to antibody activities. The second procedure using active ester–maleimide crosslinker solves the probe-orientation problem but involves pretreatments of nanoparticles and antibodies. In this approach, polymer-coated nanoparticles are first reacted with 2,2′-(ethylenedioxy) diethylamine to add a small number of amino groups and is purified with a G25 desalting column. In the mean time, purified IgG molecules are reduced by dithiothreitol to cleave the disulfide bonds in the hinge region. Similar and very detailed procedures are available at Quantum Dot Corporation's website. It has been our experiences that this antibody fragment conjugation leads to less aggregation and bioactivity retardation. Although the binding affinity of each antibody fragment to its target molecules decreases, it could be compensated by a multivalence effect (multiple fragments per QD because of its large surface area), and this matter deserves careful examination. For some applications where the whole antibody is critical for specific molecular recognition, we are developing a new conjugation chemistry based on hydrazide coupling, which not only allows the use of whole IgG, but also controls the IgG orientation. Preliminary studies have shown improved results in multicolor molecular mapping of formalin-fixed, paraffin-embedded tissue specimens.

3.3. In Vitro Cellular Imaging and Spectroscopy

PSMA-positive C4-2 cells and PSMA-negative PC-3 cells are cultured 2–3 d on chamber slides. For live cell staining, no blocking step is needed. QD–PSMA or QD–PEG bioconjugates are diluted to 50 nM in PBS or Hank's balanced buffers, and incubated (100 μL) with the cultured cells for 1 h at 4°C. The stained cells are then gently washed with PBS for three times and photographed on an inverted fluorescence microscope (Olympus, IX-70) equipped with a digital color camera (Nikon D1), a broad-band blue light source (480/40 nm, 100-W mercury lamp), and a long-pass interference filter (DM 510, Chroma Tech, Brattleboro, VT). Single cell fluorescence intensity is quantified with flow cytometer (FACS) or wavelength-resolved, single-stage spectrometer (SpectraPro 150, Roper Scientific, Trenton, NJ; detailed instrument setup is described in an early volume of this book series) *(23)*.

3.4. In Vivo Animal Imaging

3.4.1. Animal Preparation and Processing

All the protocols described next have been examined and approved by the Institutional Animal Care and Use Committee of Emory University. C4-2 prostate tumor cells are cultured 2–3 d and injected into 6–8 wk nude mice subcutaneously (10^6 cells/injection site). Tumor growth should be monitored daily until it reaches the desired sizes using caliper measurements. (For C4-2 cell line, spontaneous tumor growth varies among different animals. Therefore, each mouse is implanted with tumor cells at multiple sites.) The mice are divided into three groups for passive, active targeting, and control studies. They are then placed under anesthesia by injection of a ketamine and xylazine mixture intraperitoneally at a dosage of 95 mg/kg and 5 mg/kg, respectively. QD bioconjugates are injected through the tail vein at 0.4 nmol per mouse for active targeting, or 6.0 nmol for passive targeting and control experiment using 29X1/2-gage insulin syringes. After imaging studies, the mice are sacrificed by CO_2 overdose. Tumor and major organs (brain, heart, kidney, liver, lung, muscle, and spleen) were removed and frozen for histology examination. Tissue collections were cryosectioned into 5- to 10-μm thickness sections, fixed with acetone at 0°C, and imaged on the inverted fluorescence microscope.

3.4.2. Tumor Imaging Strategies

Bioconjugated QDs are delivered to the tumor sites by both passive and active tumor-targeting mechanisms (**Fig. 1**) *(24)*. In the passive mode, PEG-coated long-circulating QDs are accumulated preferentially at tumor sites through an enhanced permeability and retention effect; whereas in the active mode, QDs linked with targeting molecules such as antibody, peptide, antagonist, and so on quickly mark tumors through molecular recognition (in this report, QD-PSMA bioconjugate specifically bind to prostate tumors). It is worth mentioning that dextran is another attractive surface coating for enhancement of nanoparticle biocompatibility and plasma circulation time.

3.4.3. Fluorescence Imaging

In vivo fluorescence imaging is performed by using a macro-illumination system (Lightools Research, Encinitas, CA), designed specifically for small animal studies. As shown in **Fig. 2**, in a dark box illumination is provided by fiberoptic lighting (lamp house outside the dark box). For true-color fluorescence imaging, a long-pass dielectric filter (Chroma Technology) is used to reject scattered excitation light and to pass Stokes-shifted QD fluorescence. The fluorescence image is captured by a color charge-coupled device (Optronics, Magnafire SP, Olympus, America) and can be monitored on a computer screen

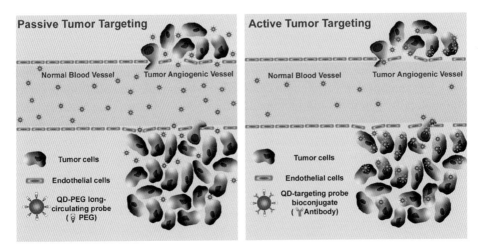

Fig. 1. In vivo tumor-targeting strategies. Passive tumor targeting based on permeation and retention of long-circulating quantum dot (QD) probes via leaky tumor vasculatures (left panel), and active tumor targeting based on high binding affinity of QD–antibody conjugates to tumor antigens (right panel).

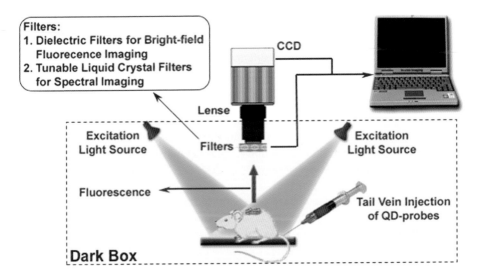

Fig. 2. Schematic illustration of in vivo optical imaging instrumentation. Tumor-bearing mice are administered with quantum dot bioconjugates intravenously and placed under anesthesia. In a dark box, illumination is provided by fiber-optic lighting. A long- pass filter is used to block the scattered lights, and a computer-controlled liquid crystal tunable filter is exploited for multispectral imaging.

in real time. For wavelength-resolved hyper spectral imaging, a cooled, scientific-grade monochrome charge-coupled device camera is used together with a spectral imaging optical head (with a built-in liquid crystal tunable filter scanning from 400 to 720 nm, CRI, Inc., Woburn, MA). Because the red QDs used in this work has an emission wavelength centered at 640 nm, the tunable filter is set to automatically step in 10-nm increments from 580 to 700 nm (*see* **Note 4**). The camera capture images at each wavelength with constant exposure, resulting in 13 TIFF images loaded into a single data structure. Based on the fluorescence spectra of pure QDs and autofluorescence, the spectral-imaging software can quickly analyze the spectral components for each pixel via a process known as "principle component analysis" *(25)*. The whole process takes less than 1 s and can be output into separate fluorescence channels or overlaid images, as shown in **Fig. 3**. It should be pointed out that the autofluorescence and QD spectra need only be recorded initially, as they can be saved in spectral libraries and reused on additional spectral unmixing.

4. Notes

1. This wavelength range provides a "clear" window for in vivo optical imaging because it is separated from the major absorption peaks of blood and water. Toward this goal, recent research has prepared alloyed semiconductor QD consisting of cadmium selenium telluride, with tunable fluorescence emission up to 850 nm *(26)*. Technical optimization of this new material together with core-shell CdTe/CdSe type-II QDs *(27)* are still needed to improve the stability and quantum efficiency. A number of promising approaches have been recently discovered. For example, Peng et al. have reported the use of successive ion layer adsorption and reaction method (originally developed for thin film deposition on solid substrates) to precisely control nanoparticle growth one layer at a time *(28)*; while Han and coworkers improved the ternary QD (three-component) synthesis by alloying the third component into preformed binary QDs *(29)*. Together with other possibilities, high-quality NIR QDs should be available very soon and bring major improvements in tissue penetration depth and cell detection sensitivity. It is worth mentioning that in vivo detection sensitivity can be further enhanced by fluorescence tomography imaging based on multiple light sources and detectors *(30)*.

2. Unmodified block copolymer can also solubilize QDs into aqueous solution, but can result in a relatively thick surface-coating layer, similar to PEG–lipid micelles *(6)*. This is because the hydrophilic methacrylic acid block doesn't have enough affinity to the particle surface and dangles in solution, which is supported by dynamic light-scattering measurements.

3. For in vivo targeting and imaging, it is advantageous to use short peptides with high binding affinity and specificity than antibodies, because of their smaller size, less immune response, lower cost, and easiness in bioconjugation. Furthermore, automated peptide synthesis and recent advance in phage display *(31)* technique enable researchers to screen and engineer peptide sequences at a relatively high throughput.

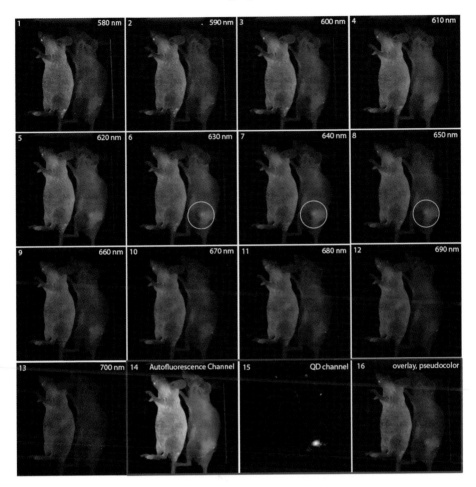

Fig. 3. Spectral imaging of quantum dot (QD)–prostate-specific membrane antigen antibody conjugates in live mice harboring C4-2 tumor xenografts. Panels 1–13, experimental raw data of an image stack from 580 to 700 nm. Please note the fluorescence intensity increase at the tumor site (white circle) from wavelength 630–650 nm, because of accumulation of red-color QDs (emission peak at 640 nm). Panels 14–16, spectrally deconvoluted images (in red square). Based on spectral distinction between mouse skin and QD emissions, fluorescent images can be output into autofluorescence and QD channels separately, or as an overlaid picture.

4. For complicated spectral deconvolution (e.g., more spectral components, similar spectra among different components, irregular spectra such as spikes, and so on), the tunable liquid crystal filter should be set to step in smaller wavelength increments, such as 1 or 5 nm. As a tradeoff of the high unmixing resolution, the imaging and computing time also increase.

Acknowledgments

This work was supported by grants from the National Institutes of Health (R01 GM60562 to SN and P01 CA098912 to LWKC), the Georgia Cancer Coalition (Distinguished Cancer Scholar Awards to both SN and LWKC), and the Coulter Translational Research Program at Georgia Tech and Emory University (to SN and LWKC). We acknowledge Professors Lily Yang, Fray F. Marshall, John A. Petros, Hyunsuk Shim, and Jonathan W. Simons for stimulating discussions and technical help. We are also grateful to Millennium Pharmaceuticals (Cambridge, MA) for providing the PSMA monoclonal antibody (J591).

References

1. Chan, W. C. W., Maxwell, D. J., Gao, X. H., Bailey, R. E., Han, M. Y., and Nie, S. M. (2002) Luminescent QDs for multiplexed biological detection and imaging. *Curr. Opin. Biotechnol.* **13,** 40–46.
2. Bruchez, M., Jr., Moronne, M., Gin, P., Weiss, S., and Alivisatos, A. P. (1998) Semiconductor nanocrystals as fluorescent biological labels. *Science* **281,** 2013–2015.
3. Chan, W. C. W. and Nie, S. M. (1998) Quantum dot bioconjugates for ultrasensitive nonisotopic detection. *Science* **281,** 2016–2018.
4. Mattoussi, H., Mauro, J. M., Goldman, E. R., et al. (2000) Self-assembly of CdSe–ZnS QDs bioconjugates using an engineered recombinant protein. *J. Am. Chem. Soc.* **122,** 12,142–12,150.
5. Akerman, M. E., Chan, W. C. W., Laakkonen, P., Bhatia, S. N., and Ruoslahti, E. (2002) Nanocrystal targeting in vivo. *Proc. Natl. Acad. Sci. USA* **99,** 12,617–12,621.
6. Dubertret, B., Skourides, P., Norris, D. J., Noireaux, V., Brivanlou, A. H., and Libchaber, A. (2002) In vivo imaging of QDs encapsulated in phospholipid micelles. *Science* **298,** 1759–1762.
7. Wu, X., Liu, H., Liu, J., et al. (2003) Immunofluorescent labeling of cancer marker Her2 and other cellular targets with semiconductor QDs. *Nat. Biotechnol.* **21,** 41–46.
8. Jaiswal, J. K., Mattoussi, H., Mauro, J. M., and Simon, S. M. (2003) Long-term multiple color imaging of live cells using quantum dot bioconjugates. *Nat. Biotechnol.* **21,** 47–51.
9. Larson, D. R., Zipfel, W. R., Williams, R. M., et al. (2003) Water-soluble quantum dots for multiphoton fluorescence imaging in vivo. *Science* **300,** 1434–1436.
10. Ishii, D., Kinbara, K., Ishida, Y., et al. (2003) Chaperonin-mediated stabilization and ATP-triggered release of semiconductor nanoparticles. *Nature* **423,** 628–632.
11. Medintz, I. L., Clapp, A. R., Mattoussi, H., Goldman, E. R, Fisher, B., and Mauro, J. M. (2003) Self-assembled nanoscale biosensors based on quantum dot FRET donors. *Nat. Mater.* **2,** 630–639.
12. Dahan, M., Levi, S., Luccardini, C., Rostaing, P., Riveau, B., and Triller, A. (2003) Diffusion dynamics of glycine receptors revealed by single–quantum dot tracking. *Science* **302,** 441–445.
13. Rosenthal, S. J., Tomlinson, I., Adkins, E. M., et al. (2002) Targeting cell surface receptors with ligand-conjugated nanocrystals. *J. Am. Chem. Soc.* **124,** 4586–4594.

14. Niemeyer, C. M. (2001) Nanoparticles, proteins, and nucleic acids: biotechnology meets materials science. *Angew. Chem. Int. Ed.* **40,** 4128–4158.
15. Alivisatos, A. P. (1996) Semiconductor clusters, nanocrystals, and quantum dots. *Science* **271,** 933–937.
16. Murray, C. B., Norris, D. J., and Bawendi, M. G. (1993) Synthesis and characterization of nearly monodisperse CdE (E = S, Se, Te) semiconductor nanocrystallites. *J. Am. Chem. Soc.* **115,** 8706–8715.
17. Hines, M. A. and Guyot-Sionnest, P. (1996) Synthesis of strongly luminescing ZnS-capped CdSe nanocrystals. *J. Phys. Chem. B* **100,** 468–471.
18. Peng, X. G., Schlamp, M. C., Kadavanich, A. V., and Alivisatos, A. P. (1997) Epitaxial growth of highly luminescent CdSe/CdS core/shell nanocrystals with photostability and electronic accessibility. *J. Am. Chem. Soc.* **119,** 7019–7029.
19. Peng, Z. A. and Peng, X. G. (2001) Formation of high-quality CdTe, CdSe, and CdS nanocrystals using CdO as precursor. *J. Am. Chem. Soc.* **123,** 183–184.
20. Qu, L., Peng, Z. A., and Peng, X. (2001) Alternative routes toward high quality CdSe nanocrystals. *Nano Lett.* **1,** 333–337.
21. Weissleder, R. (2001) A clearer vision for in vivo imaging. *Nat. Biotechnol.* **19,** 316–317.
22. Gao, X. H., Yang, L., Petros, J. A., Marshall, F. F., Simons J. W., and Nie, S. M. (2005) In vivo molecular and cellular imaging with quantum dots. *Curr. Opin. Biotech.* **16,** 63–72.
23. Gao, X. H. and Nie, S. M. (2004) Quantum dot-encoded beads. In: *NanoBiotechnology Protocols,* (Rosenthal, S. J. and Wright, D. W., eds.), Humana, Totowa, NJ, pp. 61–71.
24. Gao, X. H., Cui, Y. Y., Levenson, R. M., Chung, L. W. K., and Nie, S. M. (2004) In vivo cancer targeting and imaging with semiconductor quantum dots. *Nat. Biotechnol.* **22,** 969–976.
25. Levenson, R. M. and Hoyt, C. C. (2002) Spectral imaging in microscopy. *American Laboratory,* November, 26–33.
26. Bailey, R. E. and Nie, S. M. (2003) Alloyed semiconductor quantum dots: tuning the optical properties without changing the particle size. *J. Am. Chem. Soc.* **125,** 7100–7106.
27. Kim, S., Fisher, B., Eisler, H. J., and Bawendi, M. (2003) Type-II quantum dots: CdTe/CdSe(core/shell) and CdSe/ZinTe(core/shell) heterostructures. *J. Am. Chem. Soc.* **125,** 11,466–11,467.
28. Li, J. J., Wang Y. A., Guo, W. Z., et al. (2003) Large-scale synthesis of nearly monodisperse CdSe/CdS core/shell nanocrystals using air-stable reagents via successive ion layer adsorption and reaction. *J. Am. Chem. Soc.* **125,** 12,567–12,575.
29. Zhong, X. H., Han, M. Y., Dong, Z. L., White, T. J., and Knoll, W. (2003) Composition-tunable ZnxCd1-xSe nanocrystals with high luminescence and stability. *J. Am. Chem. Soc.* **125,** 8589–8594.
30. Ntziachristos, V., Tung, C. H., Bremer, C., and Weissleder, R. (2002) Fluorescence molecular tomography resolves protease activity in vivo. *Nat. Med.* **8,** 757–761.
31. Arap, W., Pasqualini, R., and Ruoslahti, E. (1998) Cancer treatment by targeted drug delivery to tumor vasculature. *Science* **279,** 377–380.

13

Sentinel Lymph Node Mapping
With Type-II Quantum Dots

John V. Frangioni, Sang-Wook Kim, Shunsuke Ohnishi,
Sungjee Kim, and Moungi G. Bawendi

Summary

Sentinel lymph node (SLN) mapping is an important cancer surgery during which the first lymph node draining the site of a tumor is identified, resected, and analyzed for the presence or absence of malignant cells. Fluorescent semiconductor nanocrystals (quantum dots [QDs]) of the appropriate size, charge, and emission wavelength permit this surgery to be performed rapidly, with high sensitivity and under complete image guidance. We describe the materials and methods necessary for the production and characterization of type-II near-infrared fluorescent QDs, which have been optimized for SLN mapping. They contain a CdTe core, CdSe shell, and a highly anionic, oligomeric phosphine organic coating. We also describe how to utilize such QDs in animal model systems of SLN mapping.

Key Words: Quantum dots; near-infrared fluorescence; near-infrared fluorescence imaging; imaging; intraoperative imaging; sentinel lymph nodes.

1. Introduction

In the 1990s, Morton et al. introduced the concept of sentinel lymph node (SLN) mapping and biopsy, which revolutionized the assessment of nodal status in melanoma and breast cancer *(1)*. The underlying hypothesis of SLN mapping is that the first lymph node to receive lymphatic drainage from a tumor site will contain tumor cells if there has been direct lymphatic spread *(2)*. Patients in whom SLN sampling does not reveal the presence of tumor are spared the morbidity of radical lymph node dissection.

Current techniques for SLN mapping involve preoperative injection of a radioactive colloid tracer (e.g., technetium-99m sulfur colloid) followed by intraoperative injection of a visible blue dye (e.g., isosulfan blue). The dye permits

From: *Methods in Molecular Biology, vol. 374: Quantum Dots: Applications in Biology*
Edited by: M. P. Bruchez and C. Z. Hotz © Humana Press Inc., Totowa, NJ

limited visualization of afferent lymphatic vessels and the SLN, while the radioactive colloid tracer improves detection rate and confirms complete harvest of the SLN with the use of an intraoperative hand-held γ probe *(3)*. The learning curve associated with conventional SLN mapping is steep *(4)*, the technique itself requires ionizing radiation, and the blue dye is extremely difficult to find in the presence of blood and anthracosis.

There are three important parameters in designing a lymphatic tracer for SLN mapping: hydrodynamic diameter (HD), surface charge, and contrast generation. Molecules with a HD less than approx 10 nm have the potential to travel beyond the SLN. For very small agents, such as isosulfan blue, this can result in missing the SLN, but more likely, will result in more than one nodal group in the same chain being labeled. Very large molecules, in the range of 50–100 nm, have difficulty even entering lymphatic channels, and travel so slowly that up to 24 h may be required to label the SLN (reviewed in **ref. 5**). With respect to surface charge, anionic molecules have rapid uptake into lymphatics and excellent retention in lymph nodes *(6)*. With respect to contrast generation, agents presently being used clinically are either radioactive γ emitters or colored dyes.

In 2004, our group introduced near-infrared (NIR) fluorescent quantum dots (QDs) for SLN mapping and resection. NIR light, otherwise invisible to the human eye, provides extremely high signal-to-background ratios without changing the look of the surgical field. A thorough discussion of the use of NIR light in biomedical imaging has been published previously *(7)*. When combined with a suitable intraoperative imaging system *(8,9)*, the advantage of NIR QDs for SLN mapping include high sensitivity, real-time and simultaneous visualization of both surgical anatomy and lymphatic flow, and nonradioactive detection. In this chapter, we describe in detail the production and use of type-II NIR fluorescent QDs for SLN mapping and resection. These NIR QDs have been specifically engineered with a HD (15–20 nm) that permits rapid uptake into lymphatic channels but ensures retention in the SLN, a highly anionic surface charge, maximal absorption cross-section, and a suitable quantum yield.

2. Materials

2.1. QD Chemicals

Trioctylphosphine oxide (TOPO; Alfa Aeser, Ward Hill, MA), cadmium acetylactonate (Alfa Aeser), 90% hexadecylamine (Aldrich, St. Louis, MO), 97% tri-*n*-octylphosphine (TOP; Strem, Newburyport, MA), 99.999% tellurium shot (Alfa Aeser), 98% bis-(trimethylsilyl)selenide (Acros Organics, Geel, Belgium), 90% trishydroxypropylphosphine (Strem), 98% diisocyanato-hexane (Aldrich), and 95% ethyl isocynatoacetate (Aldrich) were used as supplied. 99% Dimethylcadmium (Strem) was filtered to remove impurities using a 0.2-μm syringe filter (Pall Corporation, East Hills, NY). A 0.5 M tellurium stock solution was prepared by dissolving 3.2 g of tellurium shot in 50 mL of TOP at room temperature, stirring gently for a few hours, yielding a yellow solution.

2.2. Gel-Filtration Chromatography and Optical Measurements

The gel-filtration chromatography system consisted of an ÄKTA prime pump with fraction collector, and Superose-6 10/300 GL gel-filtration column (Amersham Biosciences, Piscataway, NJ). On-line absorbance spectrometry was performed with a model 75.3-Q-10, 1-cm path length, 70-μL quartz flow cell (Starna, Atascadero, CA), USB2000 fiber optic spectrometer, and CHEM2000-UV-VIS light source with cuvet holder (Ocean Optics, Dunedin, FL). On-line fluorescence spectrometry was performed with a model 583.4.2F-Q-10, 1-cm path length, 80-μL flow cell (Starna), HR2000 fiber optic spectrometer, CUV-ALL-UV 4-way cuvet holder (Ocean Optics), and the desired laser diode set to 5 mW with output coupled through a 300-μm core diameter, NA 0.22 fiber (Fiberguide Industries, Stirling, NJ). Laser diodes used with the system include 5W 532 nm, 5W 670 nm, or 250 mW 770 nm (Electro Optical Components, Santa Rosa, CA). Data acquisition was performed with two Dell computers using the OOIBase32 spectrometer operating software package (Ocean Optics). The spectral range of USB2000 spectrometer was from 200 to 870 nm with a spectral resolution of 1 nm, and that of HR2000 spectrometer was from 200 to 1100 nm with a spectral resolution of 6.7 nm.

2.3. Animal Models

Animals were housed in an AAALAC-certified facility staffed by full-time veterinarians, and were studied under the supervision of an approved institutional protocol. This protocol adhered to the "NIH Principles for the Utilization and Care of Vertebrate Animals Used in Testing, Research, and Training." Sprague-Dawley rats were purchased from Taconic Farms (Germantown, NY) at 300–350 g size and were of either sex. Yorkshire pigs were typically female to permit SLN mapping of mammary tissue and were purchased at 35 kg from E. M. Parsons and Sons (Hadley, MA). All animals acclimated to the animal facility for at least 48 h prior to experimentation. After each study, anesthetized rats were euthanized by intraperitoneal injection of 200 mg/kg pentobarbital, and anesthetized pigs were euthanized by rapid intravenous injection of 10 mL of Fatal-Plus (Vortech Pharmaceuticals, Dearborn, MI). These methods of euthanasia are consistent with the recommendations of the Panel on Euthanasia of the American Veterinary Medical Association.

3. Methods

3.1. Preparation of NIR Type-II QDs

Conventional type-I core-shell fluorescent semiconductor nanocrystals (QDs) confine both electrons and holes in the core. The band offsets between the core and the shell are engineered such that the conduction band of the outer-band gap material is of higher energy than that of the inner-band gap material, and the valence band of the outer-band gap material is of lower energy than that of the inner-band gap material. Quantum confinement effects in the core semi-conductor determine the range of energies possible for fluorescence in these type-I structures (**Fig. 1A**). The purpose of the shell is to provide a protective barrier that enhances the chemical stability and the quantum efficiency. In CdTe(CdSe) core(shell) QDs, the offset structure of the band implies that electrons and holes are to some extent spatially separated between the core and shell, with the holes mostly confined to the core and the electrons having a preference for the shell (**Fig. 1A**). The emission energy in these type-II structures then depends on the band offsets of the core and shell materials. Type-II QDs can therefore emit at energies smaller than either band gap of the composite materials. Both the thickness of the shell and the diameter of the core control the effective band gap of these QDs. This feature provides a new degree of flexibility in designing and engineering QD structures by largely removing the traditional correlation between QD size and its emission wavelength. This has allowed us to tailor QDs to have a size and emission wavelength specifically optimized for SLN mapping.

3.1.1. Synthesis of the Core

6.25 g TOPO, 5.75 g hexadecylamine, and 4 mL TOP were dried and degassed in the reaction vessel by heating to 140°C at approx 1 torr for 1 h and flushed periodically with argon. CdTe stock solution was prepared as follows: 634 mg (2.0 mmol) of cadmium acetylactonate was added to 6 mL of TOP and degassed at 140°C, approx 1 torr for 1 h. 4 mL of 0.5 M TOP-Te stock solution was added, mixed well, and cooled to room temperature. This solution was injected by syringe through a rubber septum at 350°C, producing a deep-red solution. The heat was then removed from the reaction vessel and the reaction left to cool to room temperature.

3.1.2. Shell Growth

CdTe QDs are precipitated from the growth solution by adding 4 mL of butanol and enough methanol to cause precipitation (~10–30 mL). Precipitated CdTe QDs (~400 mg) are collected after centrifugation (3000–6000 rpm for a few minutes), dispersed in 20 g of TOPO and 10 mL of TOP, and dried under vacuum at 160°C for 2 h. A stock solution for over-coating is prepared by mixing a 1:1 molar ratio of dimethylcadmium and bis-(trimethylsilyl)selenide in 4 mL of TOP. The amounts of precursors are calculated based on the desired shell thickness assuming epitaxial growth. While the CdTe core solution is vigorously stirred, the previously prepared overcoating stock solution is added drop-wise at 100°C for 2 h and heated to 200°C. Time and reaction temperatures vary depending on the size of CdTe QDs that are overcoated. A higher temperature is necessary to overcoat the larger CdTe QDs. The growth of the CdSe shell induces the peak red-shift of photoluminescence over time (**Fig. 1B**).

3.1.3. Oligomeric Phosphine Organic Coating

We utilize oligomeric phosphine ligands *(10)* to render the QDs stable in water (**Fig. 1C**) (*see* **Note 1**). Ligands were prepared as follows: 8.0 g of trishydroxypropylphosphine was dissolved in 20.0 g of dimethylformamide (DMF) and 4.54 g of diisocyanatohexane was added drop-wise while the solution is stirred vigorously. The reaction solution is stirred at room temperature for a day after the addition. Ethyl isocynatoacetate (19.4 g) was added drop-wise and stirred overnight. Solvent and excess ethyl isocynatoacetate were removed at 100°C *in vacuo*.

For cap exchange, 100 mg of precipitated QDs were mixed with 3.0 g oligomeric phosphine ligands in 10 mL of tetrahydrofuran and 2 mL of DMF, stirred at room temperature for 1 h, then tetrahydrofuran and DMF removed at 100°C *in vacuo*. The resultant viscous mixture was incubated at 120°C for 3 h then cooled to room temperature. 50 mL of 1 N NaOH was added, forming a two-phase suspension, and the solution was stirred vigorously at room temperature until only a single, slightly turbid dark-brown solution was present. The solution was filtered through a 0.2-μM PTFE filter (Nalgene) then ultra-filtered with 1000 vol phosphate-buffered saline (PBS), pH 7.0 using a 50-kDa cutoff membrane (Millipore).

3.2. In Vitro Characterization of Physical and Optical Properties

All QDs used in animals undergo rigorous characterization with respect to their physical and optical properties. This analysis includes the following:

3.2.1. Transmission Electron Microscopy

TEM was performed on a JEOL model 2010 electron microscope operated at 200 kV. TEM of water dispersed type-II QDs prepared as described previously showed nearly spherical core/shell dots approx 10 nm in diameter *(11)*.

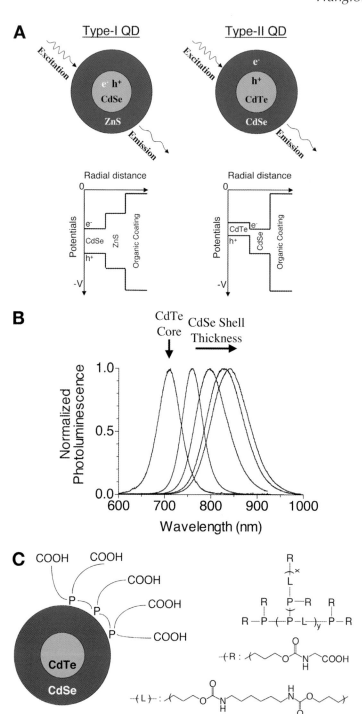

3.2.2. Estimation of QD Concentration

To obtain the extinction coefficient of NIR QDs, the total mass per QD particle was calculated by dividing the inorganic mass per particle, as measured by TEM, by the mass of inorganic cores in a dried QD sample, as measured using a Seiko model 320 thermogravimetric analyzer. The measured concentration in most cases was 1–10 μM. The concentration of QDs in solution was then obtained by Beer's law after measurement of absorption using a model HP-8453 spectrometer (Hewlett-Packard).

3.2.3. Gel-Filtration Chromatography and Optical Measurements

The gel-filtration chromatography system is shown in **Fig. 2**. Typically, 100 µL NIR QDs at a concentration of 4 µM in PBS, pH 7.8 are loaded into the injector and run at a flow rate of 1 mL/min with PBS, pH 7.8 as mobile phase. On-line, full-spectrum analysis of absorbance and fluorescence permits collection of desired fractions. Calibration of HD was performed by injecting 100 µL of a size-standard solution containing 3.8 mg/mL blue dextran (HD = 29.5 nm), 8.8 mg/mL thyroglobulin (18.8 nm), 3.8 mg/mL alcohol dehydrogenase (10.1 nm), 6.3 mg/mL ovalbumin (6.12 nm), and 2.5 mg/mL lysozyme (3.86 nm). All size standards were purchased from (Sigma). NIR QDs eluted from the column were concentrated using 10,000 MW cutoff Vivaspin (Vivascience, Edgewood, NY) concentrator with a polyethersulfone membrane (*see* **Note 2**).

This system permits real-time analysis of QD fractions and greatly simplifies the purification of QD preparations with particular optical and physical properties. For example, **Fig. 3** shows the analysis of a crude NIR QD preparation with two principle populations: one eluting at 54.13 min with a HD of 17.4 nm and another eluting at 60.27 min with a HD of 13.1 nm. As suspected from the

---

Fig. 1. Type-II quantum dots (QDs) and oligomeric phosphine organic coating. (**A**) Structure of conventional type-I QDs (left) and type-II QDs (right). The chemical compositions of core and shell are shown, as are electrons (e⁻) and holes (h⁺). Below each structure is the potential diagram for each QD layer. (**B**) A CdTe core nanocrystal was coated with an increasing thickness of a CdSe shell. Shown is the change in photoluminescence over time as the type-II QD evolves toward the desired peak emission wavelength of 840–860 nm. (**C**) An aqueous soluble organic coating on the type-II NIR QDs described in this study is formed by cap exchange of QDs with oligomeric phosphines. The monomer, trishydroxypropyl phosphine, can form both linear and branched oligomers, can be chemically exchanged with phosphines and phosphine oxides on the QD surface, and can be functionalized with carboxylic acids (this study) or other chemical substituents for control of surface charge and for conjugation to targeting ligands. The unit and oligomer structures are shown at right. The inner phosphine layer passivates the QD surface, the hydrophobic layer protects it, and the outer layer provides functionality.

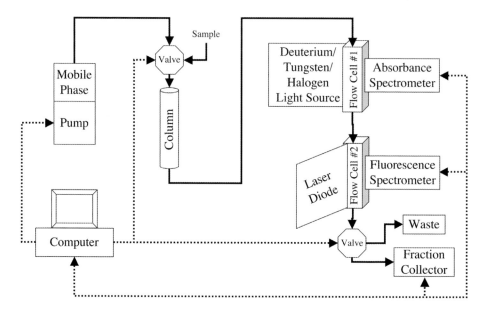

Fig. 2. On-line, full-spectrum absorbance/fluorescence gel-filtration system. Schematic diagram of the gel-filtration system components. Solid arrows denote fluid path. Dotted arrows denote computer control and data acquisition.

color of the peaks on the column (**Fig. 3B**), the absorbance and fluorescence of these populations were significantly different (**Fig. 3C**), and each peak could be collected separately for further in vitro and in vivo analysis.

3.2.4. Chemical and Photostability in Warm Plasma

To be useful for in vivo imaging, the purified NIR QD preparation must exhibit excellent chemical and photostability in warm plasma. As part of quality control for all NIR QD preparations, we measure fluorescence emission over time in warm bodily fluids *(11)*. Calf serum is used in lieu of plasma because it is more readily available, inexpensive, and has a similar protein concentration as plasma. All experiments are performed at 37°C. We specifically ensure that microaggregation is not occurring by filtering all samples through a 0.2-μm syringe filter and comparing fluorescence emission pre- and postfiltration. A typical NIR QD preparation will maintain ≥ 90% of NIR fluorescence emission, without microaggregation, after 30 min in 37°C calf serum.

3.3. SLN Mapping in Animal Model Systems

3.3.1. Animal Anesthesia

For surgery, rats were anesthetized with 75 mg/kg intraperitoneal pentobarbital. Pig anesthesia was induced with 4.4 mg/kg intramuscular Telazol (Fort

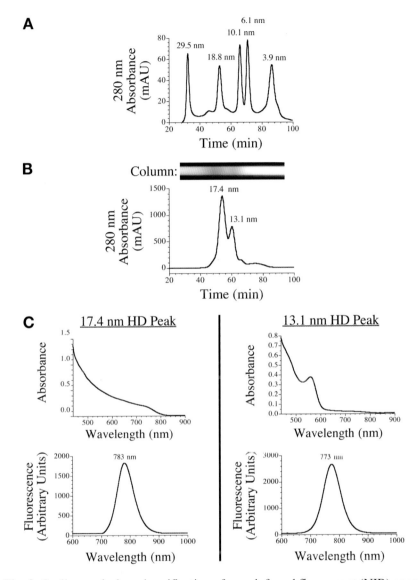

Fig. 3. On-line analysis and purification of near-infrared fluorescent (NIR) quantum dot (QD) fractions **(A)** For each NIR QD purification, calibration standards spanning HDs from 3.9 to 29.5 nm are detected using 280-nm absorbance (mAU). **(B)** The NIR QD sample is then run, with each peak calibrated in HD. Shown is the chromatogram for 280-nm absorbance (mAU). Above the chromatogram is a picture of the actual column showing the dramatic difference in color of the first (brown) and second (red) eluting peaks. **(C)** Real-time, full-spectrum absorbance (top panels) and fluorescence (bottom panels) for the 17.4-nm peak (left) and 13.1-nm peak (right). Fluorescence excitation was via a 532-nm laser diode.

Dodge Labs, Fort Dodge, IA) and anesthesia was maintained through a 7-mm endotracheal tube with 1.5% isoflurane/98.5% O_2 at 5 L/min.

3.3.2. Intraoperative NIR Fluorescence Imaging Systems

Intraoperative NIR fluorescence imaging systems optimized for small animal surgery *(8)* and large animal surgery *(9)* have been described in detail previously. Briefly, they are composed of two wavelength-isolated excitation sources, one generating 0.5 mW/cm² 400–700 nm "white" light, and the other generating 5 mW/cm² (large animal) to 50 mW/cm² (small animal) 725–775 nm light over a 8- (small animal) or 15-cm (large animal) diameter field of view. Simultaneous photon collection of color video and NIR fluorescence images is achieved with custom-designed optics that maintain separation of the white light and NIR fluorescence (>795 nm) channels. After computer-controlled (LabVIEW) camera acquisition via custom LabVIEW (National Instruments, Austin, TX) software, anatomic (white light) and functional (NIR fluorescent light) images can be displayed separately and merged. All images are refreshed up to 15 times per second. The entire apparatus is suspended on an articulated arm over the surgical field, thus permitting noninvasive and nonintrusive imaging. These imaging systems permit QDs used during SLN mapping of large animals to be detected in solid tissue up to 1-cm thick *(11)* and in lung up to 5-cm thick *(12)*.

3.3.3. QD Injection Technique

The injection technique is critical to the success of SLN mapping. Injection too shallow or too deep into tissue results in formation of a bleb, but no migration of NIR QDs from the site. Optimal technique includes keeping the needle bevel facing upward, and the syringe at a 45° angle relative to the plane of the tissue. Direct cannulation of lymphatic channels is not necessary. For pig experiments, 100 µL of NIR QDs were loaded into a 1-cc syringe equipped with a 25 gage, 0.5-in. needle. For rat experiments, 10 µL of NIR QDs were loaded into a glass Hamilton syringe equipped with a 30-gage, 0.5-in. needle. After injection, gentle thumb pressing of the injection site will increase hydrostatic pressure and greatly accelerate lymphatic flow to the SLN.

3.3.4. Real-Time SLN Mapping and Image-Guided Resection

To date, our group has utilized NIR QD technology for SLN mapping of skin *(11)*, esophagus *(13)*, lung *(12)*, gastrointestinal tract (Soltesz et al., manuscript submitted), pleural space *(14)*, and peritoneal space (Parungo et al., manuscript submitted). We now demonstrate limb mapping in the rat and mammary tissue mapping in the pig. SLN mapping of rat lower limb (**Fig. 4A**) and pig mammary tissue (**Fig. 4B**) are shown in **Fig. 4**. In both cases, the small amount and low concentration of NIR QDs injected cannot be seen on the color video images.

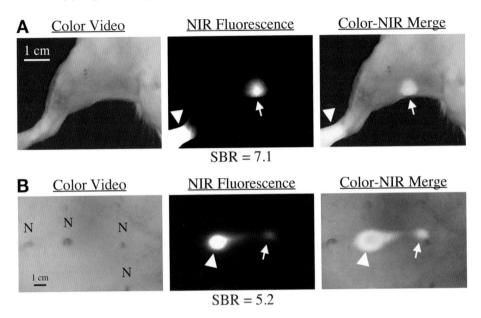

Fig. 4. In vivo sentinel lymph node (SLN) mapping in the rat and pig: Type-II near-infrared fluorescent (NIR) quantum dots (QDs) were injected (arrowhead) into rat lower limb (**A**) or pig mammary tissue (**B**), and lymphatic flow mapped in real time to the SLN (arrows). Shown are the color video image (left), the NIR fluorescence image (middle), and the pseudo-colored (lime green) merge of the two. The signal-to-background ratio of the SLN is shown below each NIR fluorescence image. NIR QD concentration was 4 μ*M*, and injected volume was 10 μL in rat and 100 μL in pig. The fluence rate of 725- to 775-nm excitation light was 5 mW/cm^2 and NIR camera exposure time was 67 ms. N, nipple.

However, NIR fluorescence imaging reveals the fine detail of lymphatic flow from the injection site to the SLN (*see* **Notes 3** and **4**). Background autofluorescence from tissue is low in this spectral region, permitting a high signal-to-background ratio to be achieved, even with 67 ms (15 Hz) exposure times (**Fig. 4**).

4. Notes

1. The selection of QD organic coating is of crucial importance. This coating is required to render the semiconductor component soluble in aqueous buffers, determines QD stability, and determines QD quantum yield in the presence of plasma proteins. It also contributes to the overall HD and controls how the particle will interact with the biological system under study. For SLN mapping, we have purposely chosen an oligomeric phosphine coating because it renders the NIR QDs highly anionic, stable in the presence of proteins, and reasonably luminescent (*11*). We also perform quality control on each sample by incubating in 37°C serum and measuring fluorescence emission and micro-aggregation over time.

2. To measure HD, we prefer gel-filtration chromatography to all other analysis methods because it has the most biological relevance. TEM is not appropriate, because only the metal component of the QD is visible. Quasi-elastic light scattering tends to weight the larger particles in a sample more than the smaller ones, thus biasing results. Gel-filtration chromatography provides both peak size and size distribution of a QD sample, and by employing on-line spectrometry, permits the desired fraction to be isolated in a single step. Most importantly, literature values for the biodistribution and pharmacokinetic behavior of biomolecules in vivo are typically derived from gel-filtration chromatography, thus permitting a direct comparison of results obtained with NIR QDs.

3. Preservation of type-II QD NIR fluorescence during histological processing has not been evaluated systematically. For most experiments, we embed QD-containing tissue in Tissue-Tek O.C.T. compound (Sakura Finetek, Torrance, CA), freeze in liquid nitrogen, and cryo-section tissue at 6–20 μm. However, there is significant loss of fluorescence during this process, most likely from damage to the nanocrystalline lattice. This is in sharp contrast to organic heptamethine indocyanine contrast agents, which exhibit full preservation of NIR fluorescence during frozen sectioning *(13,15)*. Alternative fixation and processing procedures for type-II QDs need to be explored.

4. The technology described in **Note 3** provides proof of principle experiments in large animal model systems approaching the size of humans. However, the potential toxicity of cadmium and telluride preclude their use in humans until appropriate toxicology and histopathology analyses have been conducted. Future studies should also be directed at developing NIR fluorescent QD formulations that minimize or eliminate heavy metals.

Acknowledgments

This work was supported in part by the US National Science Foundation–Materials Research Science and Engineering Center Program under grant DMR-9808941 (M. G. B.), the US Office of Naval Research (M. G. B.), the US Department of Energy (Office of Biological and Environmental Research) grant DE-FG02-01ER63188 (J. V. F.), a Proof of Principle Award from the Center for Integration of Medicine and Innovative Technology (CIMIT; JVF), and NIH grant no. R21/R33 EB-000673 (J. V. F. and M. G. B.). We thank Grisel Rivera for administrative assistance.

References

1. Morton, D. L., Wen, D. R., Wong, J. H., et al. (1992) Technical details of intraoperative lymphatic mapping for early stage melanoma. *Arch. Surg.* **127,** 392–399.
2. Cabanas, R. M. (1977) An approach for the treatment of penile carcinoma. *Cancer* **39,** 456–466.
3. Cox, C. E., Pendas, S., Cox, J. M., et al. (1988) Guidelines for sentinel node biopsy and lymphatic mapping of patients with breast cancer. *Ann Surg.* **227,** 645–653.

4. Schirrmeister, H., Kotzerke, J., Vogl, F., et al. (2004) Prospective evaluation of factors influencing success rates of sentinel node biopsy in 814 breast cancer patients. *Cancer Biother Radiopharm.* **19**, 784–790.
5. Fujii, H., Kitagawa, Y., Kitajima, M., and Kubo, A. (2004) Sentinel nodes of malignancies originating in the alimentary tract. *Ann. Nucl. Med.* **18**, 1–12.
6. Josephson, L., Mahmood, U., Wunderbaldinger, P., Tang, Y., and Weissleder, R. (2003) Pan and sentinel lymph node visualization using a near-infrared fluorescent probe. *Mol. Imaging* **2**, 18–23.
7. Lim, Y. T., Kim, S., Nakayama, A., Stott, N. E., Bawendi, M. G., and Frangioni, J. V. (2003) Selection of QD wavelengths for biomedical assays and imaging. *Mol. Imaging* **2**, 50–64.
8. Nakayama, A., del Monte, F., Hajjar, R. J., and Frangioni, J. V. (2002) Functional near-infrared fluorescence imaging for cardiac surgery and targeted gene therapy. *Mol. Imaging* **1**, 365–377.
9. De Grand, A. M. and Frangioni, J. V. (2003) An operational near-infrared fluorescence imaging system prototype for large animal surgery. *Technol. Cancer Res. Treat.* **2**, 553–562.
10. Kim, S. and Bawendi, M. G. (2003) Oligomeric ligands for luminescent and stable nanocrystal QDs. *J. Am. Chem. Soc.* **125**, 14,652–14,653.
11. Kim, S., Lim, Y. T., Soltesz, E. G., et al. (2004) Near-infrared fluorescent type II quantum dots for sentinel lymph node mapping. *Nat. Biotechnol.* **22**, 93–97.
12. Soltesz, E. G., Kim, S., Laurence, R. G., et al. (2005) Intraoperative sentinel lymph node mapping of the lung using near-infrared fluorescent quantum dots. *Ann. Thorac. Surg.* **79**, 269–277.
13. Parungo, C. P., Ohnishi, S., Kim, S. W., et al. (2004) Intraoperative identification of esophageal sentinel lymph nodes using near-infrared fluorescence imaging. *J. Thor. Cardiovasc. Surg.,* in press.
14. Parungo, C. P., Colson, Y. L., Kim, S., et al. (2005) Sentinel lymph node mapping of the pleural space. *Chest* in press.
15. Parungo, C. P., Ohnishi, S., De Grand, A. M., et al. (2004) In vivo optical imaging of pleural space drainage to lymph nodes of prognostic significance. *Ann. Surg. Oncol.* **11**, 1085–1092.

14

Macrophage-Mediated Colocalization of Quantum Dots in Experimental Glioma

Osman Muhammad, Alexandra Popescu, and Steven A. Toms

Summary

This chapter describes the methodology and a detailed protocol for the targeting and optical imaging of experimental brain tumors in rats using quantum dots (QDs). QDs are optical semiconductor nanocrystals that exhibit stable, bright fluorescence over narrow, size-tunable emission bands. The size-tunable optical properties of QDs allow multiplexing with multiple emission wavelengths from a single excitation source. QDs may be linked to antibodies, peptides, and nucleic acids for use as fluorescence probes in vitro and in vivo. We have hypothesized that the intravenous injection of QDs may represent a novel technique to optically label brain tumors, potentially leading to improved techniques for surgical biopsy and resection.

Key Words: Optical nanocrystal; nanoparticle; macrophage; brain tumor; glioblastoma; surgery; optical imaging.

1. Introduction

1.1. Optical Detection of Gliomas

The distinction of tumor from normal brain is crucial in successful brain biopsy or brain tumor resection. Despite advances in radiotherapy and chemotherapy, the clinical outcome for patients with gliomas remains poor. Consequently, surgical resection remains an important treatment modality in the care of glioma patients. Surgical resection provides both the tissue diagnosis necessary for implementation of further therapies such as radiation and chemotherapy, as well as permitting the cytoreduction of bulky neoplasms. There is growing evidence that extent of resection affects overall patient survival in gliomas *(1,2)*.

From: *Methods in Molecular Biology, vol. 374: Quantum Dots: Applications in Biology*
Edited by: M. P. Bruchez and C. Z. Hotz © Humana Press Inc., Totowa, NJ

In trying to achieve more complete glioma resections, the surgeon encounters several, persistent hurdles. Tumors are often adjacent to, or directly invade, eloquent neural structures such as motor or speech cortical structures. This increases the risk of neurological deficits during attempted resection. Techniques such as intraoperative computed tomography and magnetic resonance imaging (MRI) have shown promise in improving surgical biopsy and resection of brain tumor, but these approaches are expensive and not widely available. Optical approaches, including optical spectroscopy and the preoperative injection of fluorescent dyes, have also been used in an attempt to aid in surgical resection, but thus far are limited in their applicability because of difficulty with tissue discrimination, uptake, and photobleaching of fluorescent dyes *(3–5)*.

1.2. Optical Properties of Quantum Dots

The introduction of water-soluble quantum dots (QDs) to biological research has been successful in improving the signal-to-noise ratio in the detection of biological molecules in vitro. QDs are optical semiconductor nanocrystals that exhibit stable, bright fluorescence over narrow, size-tunable emission bands. QDs have unique optical and electronic properties: improved signal brightness, size-tunable light emission, resistance against photobleaching, and simultaneous excitation of multiple fluorescence colors in comparison with organic dyes and fluorescent proteins *(6–8)*.

QDs have an absorption spectra that exhibits an absorption peak at a wavelength slightly shorter than the QDs emission wavelength. At wavelengths shorter than this absorption peak, absorbance first declines, then rises steeply at shorter wavelengths in the blue and ultraviolet range, with no definite peak. QDs are unique among fluorophores in that they may be excited at any wavelength shorter than the emission maximum, allowing the excitation of multiple colors of QDs with a single, short-wavelength excitation source.

Work in tissue optics has shown that deep-tissue imaging is markedly enhanced by using far-red and near-infrared light in the spectral range 650 to 900 nm *(9)*. Shorter wavelengths are rapidly absorbed and scattered in tissues, preventing optical imaging at depths beyond 1 cm. We have chosen to use 705-nm Qdot nanocrystals because this wavelength provides a clear window for in vivo optical imaging at excitation wavelengths chosen for this study and would allow the optical detection of QDs within tissue several centimeters from the surface of the tissue.

Optical imaging and optical spectroscopic systems are available that can identify QDs manufactured to emit in the red or near-infrared spectra. The high quantum efficiency of Qdot nanocrystals allows an optical detection system to detect QDs in tissues by delivering an excitation light through surface light sources for optical detection by imaging systems such as charge-coupled device (CCD) cameras, as well as tissue-surface probes and biopsy needles for optical spectro-scopic systems. Detection of QDs colocalized with glioma tissue would confirm residual neoplasm is present during surgical resection and direct the surgeon to the location of the residual tumor. In the situation of brain biopsy, the detection of QDs at the tip of the biopsy probe would confirm that the needle is properly positioned prior to actual tissue biopsy. This could lead to shorter operating times, more complete surgical resections, and less morbidity in brain tumor surgery.

1.3. Water-Soluble QDs and Reticuloendothelial System Uptake In Vivo

Over the past 10 yr, techniques to make QDs water soluble have been devel-oped *(6,7)*. This has lead to a multitude of in vitro uses including nucleic acid and protein labeling for PCR and immunohistochemistry applications *(10,11)*. More recently, it has been shown that QDs may be delivered intravenously for optical angiography *(12)* and used in sentinel lymph node mapping *(13)*. Further studies of the distribution and pharmacokinetics of QDs delivered intra-venously have revealed that their circulation half-life is influenced strongly by their surface chemistry and that they are cleared from the circulation primarily by phagocytosis of the nanoparticle by the reticuloendothelial system (RES) in the liver, spleen, and lymph nodes (**Fig. 1**) *(3)*.

The primary phagocytic cell of the RES is the monocyte, or macrophage. These cells include the circulating macrophages, perivascular macrophages, and tissue macrophages. Experimental data suggests that peripheral tissue macrophages are capable of the phagocytosis of QDs (**Fig. 1D**). In the brain, a macrophage-derived cell, the microglia, is also capable of phagocytosis of nanoparticles *(14)*. Other nanoparticles, such as ultrasmall paramagnetic iron oxide particles (USPIOs), have a distribution and uptake by the RES similar to that of QDs when USPIOs are give intravenously in animals and humans *(15,16)*.

The most widely available nanoparticles with the longest history of in vivo use are the USPIOs. In reviewing the USPIO literature, it is evident that these nanoparticles are opsonized and targeted for uptake by the RES *(17)*. QDs with surface modifications that improve blood circulation half-life might allow the avoidance of the reticuloendothelial cells of the liver, spleen, and lymph nodes allowing tissue macrophages to phagocytize the circulating QDs *(18,19)*.

1.4. Nanoparticles and Tumor Targeting

Although it is possible that QDs could traverse the incomplete blood–brain barrier that is characteristic of glioma and become trapped in the interstitial spaces of gliomas, experimental data suggests that QD colocalization with glioma is mediated through the macrophage phagocytosis of QDs. Our model nanoparticle, the USPIO, has been used over the last decade in brain lesion imaging. An abundance of data exists showing that the macrophage and

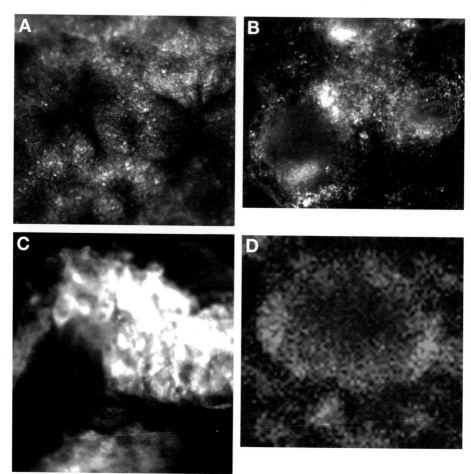

Fig. 1. After intravenous injection of quantum dots (QDs), the nanoparticles rapidly accumulate in the **(A)** liver, **(B)** spleen, and **(C)** lymph nodes (Leica stereoscope with Qdot filters and charge-coupled device camera; images ×10–20). The QDs are phagocytized by reticuloendothelial system elements including Kupfer cells of the liver **(D)**. Note the red QDs localized within the endosomes of the green CD11b-positive staining of the cell membranes of the phagocytic Kupfer cells (Leica confocal microscope, ×250).

microglia-mediated uptake of USPIOs allows these particle to be visualized within tumors and inflammatory lesions within the central nervous system *(14,20)*. In fact, macrophage-mediated delivery of USPIOs has been hypothesized as the mechanism for the delivery of magnetic particles to malignant gliomas, a technique under study for intra- and perioperative MRI of these lesions *(21)*.

These data suggest that the intravenous delivery of QDs could be accompanied by macrophage-mediated uptake and colocalization of the QD-bearing macrophages within experimental gliomas. We hypothesized that the intravenous delivery of QDs would allow the deposition of QD-laden macrophages within experimental glioma, allowing the optical detection of gliomas in animals. Optical imaging and spectroscopic techniques currently exist to use this information to guide the completeness of surgical resection of gliomas or to aid in accurate needle biopsy. Adaptation of these techniques to the surgical management of gliomas has the potential to reduce operating time, improve diagnostic accuracy, and improve patient outcomes in glioma therapy.

1.5. Potential In Vivo Toxicity of QDs

Although little toxicity has been demonstrated in animal studies of QDs, the cadmium and selenium cores have the potential for neurological and genitourinary toxicity. Current data suggests that the less protected the core or the core/shell material of the QD is, the faster the release of elemental cadmium and selenium ions, and the greater the potential toxicity *(10)*. Other semiconductor materials with less potential toxicity have been manufactured and are capable of optical emissions in the red and near-infrared range. Thus, even if a safe range does not exist for human use of QDs composed of cadmium and selenium, it is likely that other elements with semiconductor and optical properties will be available in the future for optical nanocrystal applications such as those detailed previously (**Subheading 1.2.**). Furthermore, changes in the shell that protect the QDs from nonspecific sequestration in the RES, improve circulation half-life, or more effectively target macrophages and microglia, may further decrease the total dose of QD necessary to effectively label a glioma.

2. Materials

2.1. Glioma Cell Culture

1. C6 rat glioma cells (American Type Culture Collection, Manassas, VA).
2. Phosphate-buffered saline (PBS) (Invitrogen, Carlsbad, CA).
3. 0.9% NaCl (Baxter, Deerfield, IL).
4. 1X DMEM (Invitrogen).
5. 10% Fetal bovine serum (Invitrogen).
6. Trypsin (Invitrogen).
7. Penicillin/streptomycin (Invitrogen).
8. Tissue culture flasks (Fischer, Pittsburg, PA).
9. Tissue culture incubator (Forma Scientific Model 3110, Thermo Electron Corporation, Waltham, MA).

2.2. Rat Experimental Glioma Model

1. 350-gm male Fisher rats (Charles River Labs, Wilmington, MA).
2. Ketamine (Fort Dodge Animal Health, Fort Dodge, IA).

3. Small animal stereotaxic head frame (David Kopf Instruments, Model 1900, Tujunga, CA).
4. 23-Gage syringe (Agilent Technologies, Palo Alto, CA).
5. High-speed micro drill (Fine Science Tools, Foster City, CA).
6. 3 Tesla MRI with human wrist coil (Siemans, Malvern, PA).
7. Gadolinium (Magnevist, Berlex, Montville, NJ).

2.3. QD Injection, Animal Sacrifice, and Organ Harvest

1. 705-nm Emission Qdot ITK Amino (PEG) Quantum Dots (Quantum Dot Corporation, Hayward, CA).
2. 25-Gage needle (Fisher).
3. Forceps, scissors (Roboz Surgical, Gaithersburg, MD).
4. 4% Paraformaldehyde (Fisher Chemicals, Fairlawn, NJ).
5. Sucrose (Fisher Chemicals).
6. Tissue-Tek OCT compound (Sakura Finetek, Torrance, CA).

2.4. QD Visualization in Organs

1. Xenogen IVIS system (Xenogen Corporation, Alameda, CA).
2. Leica stereoscope (Leica, Bannockburn, IL, cat. no. MZ16FA).
3. QD 705 filter set with 20-nm wide emission (Chroma, Rockingham, VT, cat. no. 32014).
4. Hamamatsu ORCA II cooled CCD camera (Hamamatsu, Bridgewater, NJ).
5. Personal computer with imaging software (Dell, Austin, TX).

2.5. High Power and Confocal Microscopy of QDs

1. Hemotoxylin (Fisher Chemicals).
2. Eosin (Fisher Scientific).
3. Monoclonal mouse antirat CD11b antibody (Serotec, Raleigh, NC).
4. Qdot 525 antimouse antibody (Quantum Dot Corp.).
5. Epifluorescence microscope (Zeiss, Thornwood, AZ).
6. Leica scanning confocal microscope (Leica, cat. no. TCS SP2).
7. Cryostat (Leica).

3. Methods
3.1. Glioma Cell Line Culture

1. C6 rat gliosarcoma cell lines are cultured at 37°C in a humidified 5% CO_2 atmosphere in DMEM supplemented with 10% fetal bovine serum and 1% penicillin/streptomycin.
2. The media is changed every 3 d.
3. The cells are passaged with tyrpsin once per week in a 1:5 ratio.
4. Cells are harvested after typsinization and washing with PBS.
5. Cells are resuspended in PBS at a concentration of 5×10^6 cells/mL prior to injection.

3.2. Rat Experimental Glioma Model

1. 350-g Male Fisher rats are anesthetized with 200 µL ketamine.
2. The animals are placed in a rat stereotactic frame.
3. The ipsilateral basal ganglia is targeted using landmarks of 2.5-mm lateral to bregma and 1-mm anterior to bregma.
4. A twist drill craniotomy is made to allow passage of the needle through the skull.
5. 1×10^6 C6 rat glioma cells are injected using a 23-gage syringe needle.
6. Rats are housed in negative-pressure isolation cages and tumors are allowed to grow for 14 d.
7. After 14 d, the rats are again anesthetized with 200 µL of ketamine and taken for MRI.
8. The animals are injected with 200 µL of gadolinium via tail vein.
9. After allowing 5 min circulation time, T1-weighted MRIs were performed on the Siemens 3T unit after positioning the animals within a human wrist coil.
10. Animals in whom tumor growth is confirmed are selected for QD injection (**Fig. 2A**).

3.3. QD Intravenous Injection

1. Rats are anesthetized with 200 µL of ketamine.
2. 1 mL of 705-nm emission Qdot ITK Amino (PEG) QDs are injected via rat tail vein over 2 min at a total concentration of 8.5 µ*M*.
3. Allow 24 h circulation time prior to repeating **steps 1** and **2** (*see* **Note 1**).
4. QDs are allowed to circulate an additional 24 h prior to animal sacrifice and organ harvesting (*see* **Note 2**).
5. Rats are sacrificed after 24 h post-QD injection with potassium chloride.
6. Brains, livers, spleens, and intraperitoneal/intraabdominal lymph nodes are harvested.
7. Half of the tissues are formalin fixed (4% paraformaldehyde, 0.1 *M* phosphate buffer) overnight, then transferred to a 20% sucrose solution for 12 h and embedded.
8. The other halves of tissues are embedded in Tissue-Tek OCT and then flash-frozen using liquid nitrogen.

3.4. QD Imaging of Organs

1. QDs are visualized in whole, fixed organs using the Xenogen IVIS system with filter set number 3.
2. Control brain, spleen, and liver (not injected with QDs) are placed into the IVIS system and an initial scanning was completed at filter set 3, with excitation time of 2 s. This initial scan is used as the background of autoflourescence of the tissues.
3. After the background scanning, sample tissues of brain, spleen, and liver collected from animals in which QDs had been injected are scanned at the same settings. After the scan the background was subtracted from the sample tissues and this scan was represented for the fluorescence of the tissues containing QDs (**Fig. 2B**) (*see* **Note 3**).
4. Stereoscopic images of thick tissue slices of the liver, spleen, lymph nodes, and brain are imaged with a Lecia stereoscope using a QD filter set (400 ± 50-nm excitation; 705 ± 20-nm emission) at magnification from ×5 to ×100.

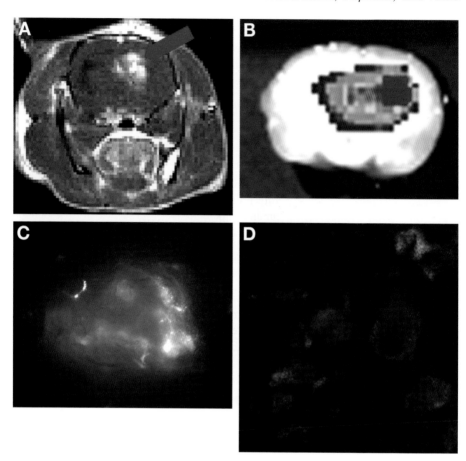

Fig. 2. MRI of rat brains 14 d after C6 glioma cell injection shows gadolinium-enhancing tumor in the rat right frontal lobe as shown by the red arrow **(A)**. Fluorescence imaging of a similar coronal section of rat brain 48 h after quantum dot (QD) injection shows intense fluorescence on both the fluoresecence-imaging systems **(B)** Xenogen IVIS system, filter set number 3, and stereomicroscopy **(C)** (Leica stereo-scope). In the brain tumor, the red QDs are confined to CD11b-positive macrophages and microglia **(D)** (Leica confocal microscope).

5. Images are collected via CCD camera and stored on a personal computer for image processing (**Fig. 2C**).

3.5. Histopathology and Tissue Confocal Microscopy

1. Both paraformaldehyde-fixed and fresh flash-frozen tissues are cryosectioned either at 9 μm (brains) or at 5 μm (livers, spleens, and lymph nodes) using a Leica cryostat.

2. The fresh flash-frozen tissue sections are then fixed in acetone for 20 min prior to immunohistochemistry.
3. Fixed tissue sections are stained with hemotoxylin and eosin and examined with bright-light microscopy for confirmation of tumor existence.
4. Fluorescence microscopy of fixed and frozen tissues is performed on an inverted epifluorescence microscope to confirm the localization of QDs. Ultraviolet filter sets allow easy excitation of all QD formulations. Because most ultraviolet microscopy systems employ long-pass filter sets, multiple colors of QDs may be visualized with this simple setting available on most microscopes to screen tissues for the presence of QDs.
5. Fresh flash-frozen tissue sections containing QDs are also examined by laser-scanning confocal microscopy for immunohistochemical analysis.
6. Macrophages are identified immunohistochemically using a primary monoclonal mouse anti-rat antibody against CD11b and a secondary Qdot 525-labeled anti-mouse antibody (**Fig. 2D**).

4. Notes

1. Initial QD injections were performed at a total concentration of 6.8 μM in a volume of 500 μL. It was determined that an increase in both total concentration and volume would provide the best chance for QD uptake by glioma by escaping early sequestration by the RES. Further increases in the injected dose of QDs to 17 nM had no apparent toxicity on the animals during the course of the experiments. At higher concentrations, optical imaging techniques were able to easily identify QDs colocalizing with the experimental gliomas at the organ and tissue slice levels.
2. Circulation times of 24–72 h allow the QDs to circulate and be taken up by macrophages, effectively colocalizing within experimental glioma. After 3 d, the number of QDs in tissues begins to decline, and signal-to-noise ratios decrease.
3. We have demonstrated that the intravenous delivery of QDs with long circulation half-life in nanomolar concentrations can escape early sequestration in the liver, spleen, and lymph nodes. Our data suggest that macrophages and microglia are efficient at phagocytizing circulating QDs. Macrophages and microglia localize within experimental glioma, optically outlining the tumor mass. These techniques have the potential to be translated into clinical use in humans, allowing QDs to optically guide brain tumor biopsies and resections.

References

1. Lacroix, M., Abi-Said, D., Fourney, D. R., et al. (2001) A multivariate analysis of 416 patients with glioblastoma multiforme: prognosis, extent of resection, and survival. *J. Neurosurg.* **95,** 190–198.
2. Toms, S. A., Ferson, D., and Sawaya, R. (1999) Basic surgical techniques in the resection of malignant gliomas. *J. Neurooncol.* **42,** 215–226.
3. Stummer, W., Novotny, A., Stepp, H., Goetz, C., Bise, K., and Reulen, H. J. (2000) Fluorescence-guided resection of glioblastoma multiforme by using 5-aminole-

vulinic acid-induced porphyrins: a prospective study in 52 consecutive patients. *J. Neurosurg.* **93**, 1003–1013.

4. Kabuto, M., Kubata, T., Kobayashi, H., et al. (1997) Experimental and clinical study of detection of glioma at surgery using fluorescent imaging by a surgical microscope after fluoroscein administration. *Neurol. Res.* **19**, 9–16.

5. Lin, W. C., Toms, S. A., Johnson, M., Jansen, E. D., and Mahadevan-Jansen, A. (2001) In vivo brain tumor demarcation using optical spectroscopy. *Photochem. Photobiol.* **73**, 396–402.

6. Chan, W. C. and Nie, S. (1998) Quantum dot bioconjugates for ultrasensitive non-isotopic detection. *Science* **281**, 2016–2018.

7. Bruchez, M., Jr., Morrone, M., Gin, P., Weis, S., and Alvisatos A., P. (1998) Semiconductor nanocrystals as fluorescent biological labels. *Science* **281**, 2013–2016.

8. Ballou, B., Lagerholm, B. C., Ernst, L. A., Bruchez, M. P., and Waggoner, A. S. (2004) Noninvasive imaging of quantum dots in mice. *Bioconjugate Chem.* **15**, 79–86.

9. West, J. L. and Halas, N. J. (2003) Engineered nanomaterials for biophotonics applications: improving sensing, imaging, and therapeutics. *Annu. Rev. Biomed. Eng.* **5**, 285–292.

10. Michalet, X., Pinaud, F. F., Bentolilia, L. A., et al. (2005) Quantum dots for live cells, in vivo imaging, and diagnostics. *Science* **307**, 538–544.

11. Gao, X., Yang, L., Petros, J. A., Marshall, F. F., Simons, J. W., and Nie, S. (2005) In vivo molecular and cellular imaging with quantum dots. *Curr. Opin. Biotechnol.* **16**, 63–72.

12. Larson, D. R., Zipfel, W. R., Williams, R. M., et al. (2003) Water-soluble quantum dots for multiphoton fluorescence imaging in vivo. *Science* **300**, 1434–1436.

13. Kim, S., Lin, Y. T., Soltesz, E. G., et al. (2004) Near-infrared fluorescent type II quantum dots for sentinel lymph node mapping. *Nat. Biotechnol.* **22**, 93–97.

14. Corot, C., Petry, K. G., Trivedi, R., et al. (2004) Macrophage imaging in the central nervous system and in carotid atherosclerotic plaque using ultrasmall superparamagnetic iron oxide in magnetic resonance imaging. *Invest. Radiol.* **39**, 619–625.

15. Vassallo, P., Matei, C., Heston, W. D., McLachlan, S. J., Koutcher, J. A., and Castellino, R. A. (1994) AMI-227-enhanced MR lymphangography: usefulness for differentiating reactive from tumor-bearing lymph nodes. *Radiology* **193**, 501–506.

16. Dupas, B., Berreur, M., Rohanizadeh, R., Bonnemain, B., Meflah, K., and Prada, G. (1999) Electron microscopy study of intrahepatic ultrasmall superparamagnetic iron oxide kinetics in the rat. Relationship with magnetic resonance imaging. *Bio. Cell* **91**, 195–208.

17. Kircher, M. F., Mahmood, U., King, R. S., Weissleder, R., and Josephson, L. (2003) A multimodal nanoparticle for preoperative magnetic resonance imaging and intra-operative optical brain tumor delineation. *Ca. Res.* **63**, 8122–8125.

18. Zhang, Y., Kohler, N., and Zhang, M. (2002) Surface modifications of superparamagnetic magnetite nanoparticles and their intracellular uptake. *Biomaterials* **23**, 1553–1561.

19. Moghimi, S. M. and Szebeni, J. (2003) Stealth liposomes and long circulating nanoparticles: critical issues in pharmacokinetics, opsonization, and protein-binding properties. *Prog. Lipid Res.* **42**, 463–478.

20. Fleige, G., Nolte, C., Synowitz, M., Seeberger, F., Kettenmann, H., and Zimmer, C. (2001) Magnetic labeling of activated microglia in experimental glioma. *Neoplasia* **3,** 489–499.
21. Hunt, M. A., Bago, A. G., and Neuwelt, E. A. (2005) Single dose contrast agent for intraoperative MR imaging of intrinsic brain tumors by using ferumoxtran-10. *AJNR Am. J. Neuroradiol.* **26,** 1084–1088.

IV

FLOW CYTOMETRY APPLICATIONS

15

Application of Quantum Dots to Multicolor Flow Cytometry

Pratip K. Chattopadhyay, Joanne Yu, and Mario Roederer

Summary

The coupling of quantum dots (QDs) to antibodies heralds a new era of versatility in multicolor flow cytometry. This chapter introduces the properties of QDs (as relevant to flow cytometric applications), the advantages derived from these properties, and the procedure for conjugating antibodies to QDs. Finally, we discuss strategies for choosing the combinations of QDs and antibodies best suited for multicolor experiments.

Key Words: Quantum dots; flow cytometry; antibody conjugation; fluorochrome; multicolor; compensation.

1. Introduction
1.1. Characteristics of Quantum Dots

The fluorescent properties of quantum dots (QDs) are derived from their nanocrystal cores, which are synthesized from cadmium selenide (for QDs emitting light in the 525- to 655-nm range) or cadmium telluride (for QDs emitting higher wavelength light). The size of each particle is the primary determinant of its emission spectrum, as the smallest QDs (3 nm) emit light in the blue region of the spectrum, whereas the largest QDs (approx 6 nm) emit far red light *(1)*. Currently, nine types of QDs are available (each named according to the wavelength at which they emit maximum light). Although they fluoresce at different wavelengths, they are excited at the same wavelength, allowing detection of multiple QD colors from just one laser (*see* **Table 1**; **Subheading 1.3.**).

QDs are particularly well suited to multicolor applications of flow cytometry because each has a narrow emission spectrum that overlaps only minimally with the other QDs *(2)*. This characteristic reduces the spillover signal from other QDs

From: *Methods in Molecular Biology, vol. 374: Quantum Dots: Applications in Biology*
Edited by: M. P. Bruchez and C. Z. Hotz © Humana Press Inc., Totowa, NJ

Table 1
Quantum Dots Available for Flow Cytometry

QD	Dichroic filter	Bandpass filter	Maximum emission, color
800			800, far red
705	685	705/70	705, far red
655	630	660/40	655, far red
605	595	605/40	605, red
585	575	585/42	585, yellow
565	550	560/40	565, green
545	535	550/40	545, green
525	505	515/20	525, blue

that must be subtracted (i.e., compensated) from the primary QD channel. In practice, we find that five QDs can be used simultaneously with only minimal (<10%) compensation between channels *(2)*. Moreover, when QD reagents are used with common fluorochromes excited by 488, 532, or 633 lasers (e.g., fluoroscein isothiacyanate, phycoerythrin [PE], or allophycocyanin), we find almost no spillover signal from other fluorochromes in the QD channels *(2)*. Thus, in instruments with two or more lasers, QDs can be multiplexed with other fluorochromes to successfully measure even more colors.

However, to be useful for flow cytometry applications, the intrinsically unstable and highly reactive QD cores must be modified significantly. To this end, QD cores are stabilized with a coating of inorganic zinc-sulfide (the shell), which is in turn coated with organic polymers *(1)*. These organic polymers increase solubility and provide a platform of functional groups (such as amines, NH_3) for conjugation to antibodies, streptavidin, and nucleic acids. Because they have similar coatings, QDs of various colors share uniform biophysical properties and a common conjugation procedure.

1.2. Conjugation of QDs to Antibodies

The procedure for conjugation of antibodies to QDs is similar to conjugation of antibodies to PE, with slight variations in the reagents used and ratio of antibodies to fluorescent molecules. Successful conjugation relies on the coupling of malemide groups on the QDs to thiol groups on the antibody. These groups are generated during the initial steps of the procedure, as amine groups on the QDs are activated with a heterobifunctional crosslinker (sulfosuccinimidyl 4-[*N*-maleimidomethyl]cyclohexane-1-carboxylate, sulfo-SMCC) to generate the malemide moieties, and disulfide bonds in the antibody are reduced to thiol groups using dithiothreitol (DTT). Before conjugation, the DTT-reduced antibody is then mixed with two dye-labeled markers, Cyanin-3 (Cy3) and dextran blue, which track the monomeric fraction of antibodies as it passes through purification columns. Activated QDs and reduced antibody are subsequently purified over columns and mixed for conjugation.

1.3. Detection of QDs and Selection of QDs/Antibody Combinations

Addition of QDs to the repertoire of flow cytometry reagents has significantly increased the number of markers we can study simultaneously, thereby allowing great versatility in experimental design. Although QD technology has been relatively easy to implement, adherence to some detection and panel design guidelines increases the likelihood of success. In the procedure that follows (**Subheading 3.2.**), we detail the considerations involved in instrument configuration and in designing staining panels that use QD reagents. A protocol is also provided for the titration of QD-conjugated antibodies and for the labeling cells with titrated antibodies.

2. Materials

2.1. Conjugation of QDs to Antibodies

2.1.1. Unconjugated Antibodies and QDs

1. Unconjugated antibodies are generally obtained from commercial vendors and stored at 4°C or frozen. Frozen materials should be centrifuged prior to use (e.g., at 13,000*g* in a microcentrifuge) to remove aggregates. Unconjugated antibodies should be stored at high concentrations (generally more than 1 mg/mL, usually approx 10 mg/mL).
2. QDs are obtained from Quantum Dot Corporation (Hayward, CA), prefabricated with the organic shell (PEG) and amine groups for conjugation. Store at 4°C; do not freeze.

2.1.2. Derivitization (Activation) of QDs

1. 100 m*M* Sulfo-SMCC (Pierce, Rockford, IL; stored at 4°C in a dessicator) solution in anhydrous dimethyl sulfoxide (DMSO) (Sigma, St. Louis, MO; stored in a dessicator at room temperature). Discard after use (*see* **Note 1**).
2. Exchange buffer: 20 m*M* 2-[*N*-morpholino] ethane sulfonic acid (MES), 2 m*M* EDTA (*see* **Note 2**).

2.1.3. Reduction of Antibody

1. 1 *M* DTT (stored at 4°C in a dessicator) solution in distilled water. Discard after use (*see* **Note 3**).
2. 100X Cy3 (20 mg/mL) and 25X dextran blue (50 mg/mL). Both may be stored at 4°C indefinitely.

2.1.4. Conjugation

1. 10 mg/mL *N*-ethylmaleimide (Sigma) solution in anhydrous DMSO. Discard after use.

2. Storage buffer: 10 m*M* TRIZMA 8.0 (Sigma), 150 m*M* NaCl (*see* **Note 4**).

2.2. Detection of QDs and Selection of QDs/Antibody Combinations

1. We use a custom-configured LSRII flow cytometer (BD, San Jose, CA), equipped with a 408-nm laser for excitation of QDs (and 488, 532, and 633-nm lasers for detection of other fluorochromes) *(3)*.
2. Filters for detection of QDs (listed in **Table 1**) may be purchased from (Chroma Technologies, Rockingham, VT).

2.3. Staining Cells with QD-Labeled Antibodies

1. Staining media: biotin-free RPMI-1640 or DMEM-complete media (Gibco, Grand Island, NY; *see* **Note 5**) is generally used. Staining media is supplemented with 1% heat-inactivated fetal calf serum to ensure cell viability during staining, and 0.02% sodium azide to prevent bacterial growth during storage (*see* **Note 6**). Staining media may be stored at 4°C for 6 mo. Alternatively, cells may be stained in phosphatebuffered saline (Gibco).
2. Fixative: dilute 20% paraformaldehyde stock solution (PFA; Tousimis Research Corporation, Rockville, MD) in staining media or phosphate-buffered saline to a working concentration of 1% PFA. 1% PFA solution is stable at 4°C for 6 mo (*see* **Note 7**).
3. Cell suspension: in general, 5×10^6 (or fewer) cells are stained per test. For each test, resuspend cells in a volume of 50 µL staining media.
4. Compensation: anti-mouse Igκ beads (BD) stained with the appropriate antibodies. Software compensation was performed in FlowJo v6.2.1 (Tree Star, Inc., Ashland, OR).

3. Methods
3.1. Conjugation of QDs to Antibodies
3.1.1. Preparation of Unconjugated Antibodies and QDs

1. Calculate the moles of antibody to conjugate (*see* **Note 8**).
2. Determine the moles of QD to conjugate from the molarity of the QD material (*see* **Note 9**).
3. QD will be reacted with antibody at 1:1 ratio. Calculate the volume of QD needed for each antibody (*see* **Note 10**) by dividing the moles of antibody by the mol/mL (molarity) of the QD material.

3.1.2. Derivitization (Activation) of QDs

1. To each QD reagent, add 1% sulfo-SMCC (*see* **Note 11**).
2. Place on rotator at room temperature for 1 h.
3. During this incubation, obtain and label one NAP-5 or PD-10 column for each QD reagent or antibody (*see* **Note 12**). Equilibrate columns with five passages of exchange buffer.

3.1.3. Reduction of Antibody

1. For every 100 µL of unconjugated antibody, add 2 µL of 1 *M* DTT solution. Incubate for 15 min at room temperature.
2. During reduction, obtain 25X dextran blue and 100X Cy3 and warm to room temperature.
3. After reduction, antibodies with less than 250 µL should be resuspended to this volume in a solution of 1/25th volumes blue dextran, 1/100th volumes Cy3, and MES (*see* **Notes 13–15**).
4. Pass each derivitized QD or reduced antibody over the appropriate, MES-equilibrated column. Collect only the blue (not the pink) fraction from each column (*see* **Note 16**).

3.1.4. Conjugation

1. Mix each antibody with the appropriate QD for conjugation. Carry out the conjugation on freshly collected antibodies.
2. Incubate antibody–QD mixtures on rotator (at room temperature) for 1 h.
3. To block the unreacted sulfhydryl groups on the IgG, add 3 µL *N*-ethylmaleimide to each conjugate. Incubate 20 min.
4. During incubation, obtain and label one NAP-5 or PD10 column for each conjugate (*see* **Note 17**). Equilibrate columns with five passages of storage buffer.
5. Spin conjugates for at least 3 min at maximum speed. Transfer aqueous fraction into new tubes.
6. Add conjugates to columns, and collect colored fraction of each conjugate into storage vials (*see* **Notes 18** and **19**).

3.2. Selection of QD/Antibody Combinations, Staining Procedure, and Instrument Configuration

3.2.1. Guidelines for Selecting Antibody–QD Combinations

1. Blue (525nm) and green (545, 565nm) QDs give the dimmest signals, while redder QDs (particularly QD 655) are the brightest.
2. Thus, antibodies that stain markers brightly, with distinct separation between the positives and the negatives, should be assigned to the dim channels (525, 545, 565).
3. However, the compensation requirement between 545 and 565 can be high, and severe spreading of the negative populations in each is often observed (*see* **Note 20**). Therefore, if these QDs must be used together, reserve them for antibodies that are very well separated. Alternatively, QD 545 and 565 can be employed for markers that are not usually coexpressed on the same cells (e.g., CD4 and CD8).
4. Although signals from non-QD reagents do not generally spill into QD channels, signals from QDs can require significant compensation from non-QD channels. For example, QDs 605 and 655 emission spill into Texas Red-PE (TRPE) and Cy5PE. However, compensation of this spillover is possible, and the bright TRPE and Cy5PE signals are usually unaffected by compensation-induced spreading

of negative populations. Nevertheless, when designing and testing a panel, it is worthwhile to try the panel without the potentially problematic QD (a "fluorescence minus one" [FMO] control, *see* **Note 21**) to make sure that the resolution of the marker in the non-QD channel is unaffected.

3.2.2. Staining Cells With QD-Labeled Antibodies

1. After conjugation, and before experimental use, all QD-labeled antibodies should be titrated to determine the optimal staining concentration. To this end, cells are stained (as described in **steps 2–8**) with varying concentrations of QD-labeled antibody, and the antibody concentration yielding the brightest specific signal, with the least nonspecific staining, is subsequently used (*see* **Notes 22 and 23**). For example, cells stained with 5 μL QD655-anti-CD45RA (per 100-μL staining volume; **Fig. 1A**, top row) have high levels of nonspecific (CD45RA-) staining. This nonspecific signal dramatically decreases with antibody concentration, reaching minimal levels at 0.15 μL reagent. At this concentration, specific (CD45RA+) signal is still bright and easily resolved; thus, in future experiments, 0.15 μL QD655-anti-CD45RA will be used. For QD705-anti-CD3 (**Fig. 1A**, bottom row), resolution of CD3+ populations declines significantly beyond the optimal concentration of 2.5 μL reagent per 100 μL staining media.
2. In general, cell staining is performed by mixing 50 μL of cell suspension with a 50 μL solution of the antibodies selected for study. This yields a total staining volume of 100 μL (*see* **Note 24**).
3. To prepare the 50-μL solution of antibodies, determine the volume of each QD-labeled antibody that represents the optimal staining concentration, and transfer to a microcentrifuge tube. For example, to stain cells with the titrated reagents described in **step 1**, prepare a tube containing 0.15 μL QD655-anti-CD45RA and 2.5 μL QD705-anti-CD3. Bring antibody solution to a volume of 50 μL by adding 47.35 μL staining media.
4. When simultaneously staining with multiple conjugates, aggregates of antibodies and fluorochromes may form. To clean antibody solution, spin briefly in microcentrifuge at maximum speed. Transfer supernatant to fresh tube.
5. Add antibody solution to cell suspension. Mix thoroughly using pipet.
6. Incubate samples in the dark, at room temperature, for 15 min (*see* **Note 25**).
7. Wash cells by resuspending in 1 mL staining media. Centrifuge for 5 min at approx 860*g*, and aspirate the supernatant. Repeat.
8. Resuspend stained cells in 200 μL staining media. Add 200 μL of 1% working PFA solution (final PFA concentration, 0.5%). Mix thoroughly with pipet.
9. To adjust for the spectral overlap between QDs and non-QD fluorochromes, compensation specimens are prepared as follows. Label a separate test tube for each antibody studied, then aliquot 40 μL of Igκ beads into each. Add the appropriate antibody to each tube, incubate at room temperature (in the dark) for 15 min. Resuspend beads in 200 μL staining media, add 200 μL 1% working PFA solution.

3.2.3. Instrument Configuration

1. QDs are excited by all short-wavelength lasers in a flow cytometer. However, the brightest signal is obtained from the 355-nm laser, followed by the 405 and 488-nm lasers *(2)*. Note that even though QDs are excited by multiple lasers, the spectral spillover from QDs into other fluorochromes can be compensated.

2. Narrow dichroic filters pass emitted light between detectors, whereas wide bandpass filters are used in front of each detector to collect maximum light from a specific QD. The filters we use are listed in **Table 1**, and the instrument configuration is further described in **ref. 3**.

3. Although there is a slight spillover of emitted light between neighboring QDs with the spectrums of QD545 and 565 overlapping the most, alternate dots have almost no spillover; thus, in systems with only a few detectors, the configuration requiring the least compensation uses QDs spaced well apart on the spectrum (e.g., 525, 605, 705). Our instrumentation successfully employs eight detectors off of the 408-nm laser, so that seven QDs and Cascade or Pacific Blue can be measured simultaneously *(3)*.

4. An example of successful antibody selection, staining, compensation, and analysis with QD and non-QD fluorochromes is provided in **Fig. 1B**. All cell subsets (CD3, CD4, CD8, naïve, central memory, effector memory) were distinct and well resolved when cells were stained with QD705-anti-CD3, PE-anti-CD4, QD605-anti-CD8, QD655-anti-CD45RA, and Pacific Blue-anti-CCR7.

4. Notes

1. Use sulfo-SMCC, not SMCC. The molecular weight of sulfo-SMCC is 436.37 g/mol; therefore, 100 mM sulfo-SMCC can be prepared by dissolving 2.18 mg sulfo-SMCC in 50 µL anhydrous DMSO.

2. Exchange buffer is generally prepared in bulk, using 9.76 g MES and 4 mL of 0.5 M EDTA resuspended with distilled water to total volume of 1 L. Adjust pH to 6.0. Store at 4°C for up to 1 yr.

3. The molecular weight of DTT is 154.2 g/mol; therefore, 1 M solution can be prepared by dissolving 15.4 mg DTT in 100 µL of distilled water.

4. Storage buffer is generally prepared in bulk, stored at 4°C. Combine 1.42 g of TRIZMA 8.0, 8.77 g of NaCl, and 5 mL of a 20% sodium azide solution. Resuspend to 1 L with distilled water, and adjust pH to 8.2.

5. Biotin-free RPMI-1640 and DMEM-complete media is obtained from vendor by special order.

6. To limit toxicity at physiological temperatures, cell staining at 37°C is performed in media lacking sodium azide.

7. Do not use 1% PFA solution for more than 6 mo. Expired PFA may break down to formic acid, which decreases fluorescence intensity of some fluorochromes.

8. Assume that 1 mol of antibody = 160,000 Daltons (or 160 kDa).
   Moles of antibody = grams of antibody/160,000
   (For example, 1 mg of antibody = 6.3×10^{-9} mol)

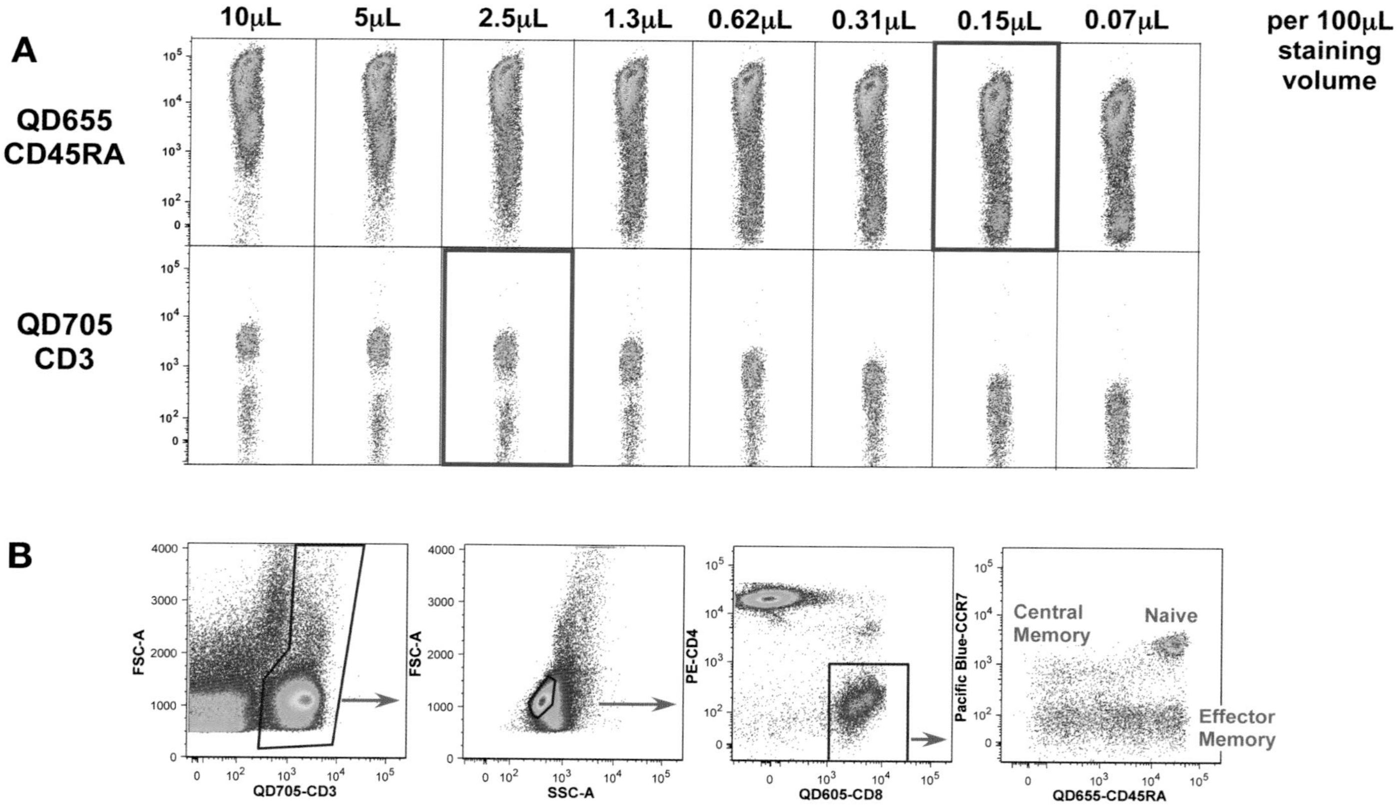
A
B
10μL
5μL
2.5μL
1.3μL
0.62μL
0.31μL
0.15μL
0.07μL
per 100μL staining volume
QD655 CD45RA
QD705 CD3
FSC-A
QD705-CD3
SSC-A
PE-CD4
QD605-CD8
Pacific Blue-CCR7
QD655-CD45RA
Central Memory
Naive
Effector Memory

Fig. 1.

9. For example, QD 605-PEG-NH2 was supplied at 8.7 μ*M*, therefore 1 mL = 8.7 × 10^{-9} mol of QD.

10. For example, if 6.3 × 10^{-9} mol (i.e., 1 mg) antibody are reacted with 6.3 × 10^{-9} moles of QD (1:1 antibody:QD ratio), then set aside 6.3 nmol/8.7 nmol/mL (QD molarity), or 724 μL of QD reagent for conjugation.

11. For example, when conjugating 724 μL QD, add 7.24 μL of 100 m*M* sulfo-SMCC.

12. Use PD-10 column for volumes over 0.5 mL.

13. For example, for 50 μL of antibody solution, 200 μL of buffer will be added. The buffer should be premade with dextran blue and Cy3, e.g., 250-μL total volume/ 25 = 10 μL of 25X dextran blue to the MES buffer prep and 2.5 μL Cy3.

14. Dextran blue is very viscous. To pipet, first coat the tip in dextran blue by pipetting up and down repeatedly.

15. When adding dextran blue to antibody, pipet up and down to mix, but take care not to generate bubbles, which can "reoxidize" the IgG.

16. Dextran blue tracks with the monomeric antibody fraction (i.e., 160 kDa), whereas the larger Cy3 tracker (pink) tracks with higher order antibody complexes and aggregates.

17. PD10, not NAP-5, columns are usually used at this step.

18. QDs may adhere to the plastic surface of untreated microcentrifuge tubes; therefore, we generally collect and store antibody conjugates in glass vials.

19. Some antibodies do not work well with QDs for as-yet-unknown reasons. Thus, all newly conjugated materials should be titrated and validated against commercial (non-QD) reagents.

20. Compensation-related spreading is discussed in **ref. 4**. Briefly, when photons of emitted light are enumerated by a flow cytometer, errors in the counting-statistics inevitably arise. These errors are revealed upon compensation as a spreading of the negative population in the compensated channel. For fluorochromes that overlap significantly, this spreading can be dramatic and causes difficulties in resolving dimly staining populations.

Fig. 1. (A) Titration analyses for QD655-anti-CD45RA and QD705-anti-CD3. Cells were stained with varying concentrations of quantum dot (QD)-labeled antibodies. The concentrations with the least nonspecific (CD45RA- or CD3-) staining and the brightest specific (CD45RA+ or CD3+) signal are indicated in red for each antibody. These optimal staining concentrations (indicated in red) were chosen for subsequent experiments. Titration analysis should be performed for every new conjugate, and when experimental conditions (such as staining temperature) are altered. (B) Successful staining and analysis using QD and non-QD fluorochromes. Cells were stained with the following panel of titrated antibodies: QD705-anti-CD3, PE-anti-CD4, QD605-anti-CD8, QD655-anti-CD45RA, and Pacific Blue-anti-CCR7. CD4+ and CD8+ T-cells were easily distinguished, as were naïve (CD45RA+ CCR7+), central memory (CD45RA- CCR7+), and effector memory (CCR7-) subpopulations.

21. The effect of error-induced spreading can be determined using FMO controls *(3)*. FMO controls contain all the antibodies of interest, except one. In the scenario described in the protocol, the FMO control would lack QD655. Staining in the TRPE channel would be compared between the FMO control and the complete panel, to ensure that all populations were adequately resolved.
22. Lower staining concentrations (i.e., below the optimum) may be used in cases where the nonspecific signal remains well separated from specific signal.
23. Titrations should be performed under the same experimental conditions that will be used in subsequent experiments. When staining conditions (such as incubation temperature or target cell type) are altered, titrations should be repeated. Also, optimal concentrations may differ between reagents, and even from lot-to-lot for the same reagent. Therefore, titrations should be performed after each conjugation.
24. When staining large numbers of cells, increase the total staining volume. For example, to stain 100×10^6 cells with QD655-CD45RA, use a total staining volume of 300 µL. This would require the addition of 3X more QD655-CD45RA (i.e., 0.45 µL).
25. The measurement of some markers may require alternate incubation times and temperatures. Staining with QD-labeled antibodies has been successful at a variety of temperatures, from 4 to 37°C.

References

1. Bruchez, M., Moronne, M., Gin, P., Weiss, S., and Alivisatos, A. P. (1998) Semiconductor nanocrystals as fluorescent biological labels. *Science* **281,** 2013–2016.
2. Roederer, M., Perfetto, S. P., Chattopadhyay, P. K., Harper, T., and Bruchez, M. (2004) Quantum dots for multicolor flow cytometry. Poster, International Society for Analytical Cytology Conference; May 24–28, Montpelier, France.
3. Perefetto, S. P., Chattopadhyay, P. K., and Roederer, M. (2004) Seventeen-colour flow cytometry: unravelling the immune system. *Nat. Rev. Immun.* **4,** 648–655.
4. Roederer, M. (2001) Spectral compensation for flow cytometry: visualization artifacts, limitations, and caveats. *Cytometry* **45,** 194–205.

16

Quantum Dots in Flow Cytometry

Barnaby Abrams and Tim Dubrovsky

Summary

The development of new fluorophores has experienced a tremendous advance over the last two decades. The unique photophysical properties of quantum dots (QDs), such as their large Stokes shifts and exceptional brightness, make them attractive probes in flow cytometry applications. In this chapter, the spectral overlap and the fluorescence intensity of a single Qdot nanocrystal species (Qdot-655) was investigated in the context of a panel containing conventional fluorophores. Certain compensation issues remain because of the unique absorption characteristics of QDs. To demonstrate the potential of QDs for multi-color flow cytometry, human lymphocytes were surface stained with an eight-color panel where one of its standard violet laser reagents, CD4 AmCyan, was substituted with the CD4 Qdot-655 conjugate.

Key Words: Multicolor flow cytometry; quantum dots; spectral overlap; compensation.

1. Introduction
1.1. Historical Background

The first systematic studies of size-dependent optical properties of semiconductor crystals in colloidal solutions were performed two decades ago by Henglein *(1)* and Brus *(2)*. A major advance in increasing the quantum yield was accomplished by introducing the passivation shell on the crystalline core. The large surface-to-volume ratio in a nanosized crystal (about 50% of all atoms are on the surface) affects the emission of photons. The shell helps to confine the excitation to the CdSe core and prevent the non-radiative relaxation. Photochemical oxidation and surface defects in a crystal with no shell may lead to a broad emission and lower quantum yields. One of the first core–shell syntheses was performed by Spanhel et al. *(3)*. Major improvements leading to highly fluorescent QDs were made in the mid-1990s *(4–6)*. Subsequently, CdSe

From: *Methods in Molecular Biology, vol. 374: Quantum Dots: Applications in Biology*
Edited by: M. P. Bruchez and C. Z. Hotz © Humana Press Inc., Totowa, NJ

crystals with silane-modified hydrophilic surfaces were introduced as reporters for biological applications *(7,8)*. Even more recent developments include the encapsulation of CdSe/ZnS core–hell nanocrystals into carboxylated polymer, followed by chemical modification of the surface with long-chain polyethylene glycol (PEG) (Quantum Dot Corporation, Hayward, CA).

1.2. Unique Photophysical Properties of Quantum Dots

Fluorescence in CdSe semiconductor crystals is because of the radiative recombination of an excited electron–hole pair, or exciton. The excitonic energy levels and quantum yields of fluorescence, i.e., the ratios describing how many absorbed photons are emitted, depend on exciton–photon interaction in the crystal and the size of the crystalline core. Each quantum dot (QD) has its characteristic singlet emission peak. With increasing crystal size (from 2–3 to 10–12 nm), the emission maximum shifts from 500 to 800 nm. In addition, QDs have very broad absorption spectra, and can be excited over the entire visual wavelength range as well as far into the ultraviolet (*see* **Fig. 1**). Because of their exceptionally large Stokes shifts (up to 400 nm), QDs can potentially be used for the multicolor detection with a single wavelength excitation source.

1.3. Fluorescence Intensity and Lifetime

The core–shell nanocrystals have large extinction coefficients and high quantum yields. These parameters describe the capacity of the system to capture and subsequently rerelease light. Although quantum yields of Qdot conjugates in aqueous buffers (20–50%) are comparable with those of conventional fluorophores, the excitation efficiency of Qdot conjugates is much higher. The CD4 Qdot conjugate has an absorption maximum of $6 \times 10^6\ M^{-1}cm^{-1}$ at 405 nm (*see* **Fig. 1**). This makes these Qdot conjugates about two orders of magnitude more efficient at absorbing excitation light than organic dyes and fluorescent proteins.

QDs have a fluorescence lifetime of 20–30 ns—about 10 times longer than the background autofluorescence of proteins. The lifetime characterizes a delay between the moment in which the QD absorbs a photon from the light source and the moment of radiative recombination of the exciton. Fluorescence from single CdSe crystals has been observed much longer than from other fluorophores, resulting in high turnover rates and a large number of emitted photons *(9)*. When excited at 405 nm, CD4 Qdot-655 conjugates are about 80–100 times brighter than CD4 AmCyan and CD4 Pacific Blue conjugates (*see* **Fig. 2**).

1.4. Recent Developments in Surface Chemistry

Two important obstacles to biological applications of commercially available QDs were low quantum yields in aqueous buffers and strong aggregation of conjugates. Surface chemistry played a major role *(10)*. For the use of QDs as

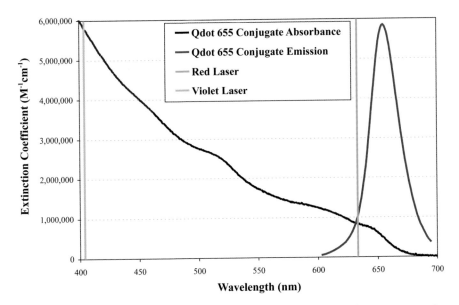

Fig. 1. Absorption and emission spectra of Qdot-655 conjugate. Bars represent laser lines of the violet and red lasers. The shape of the absorption spectrum reveals that the QD can be excited by several commonly used visible-range emission lasers (e.g., violet, blue, green, and red). Although the extinction coefficient is much lower at the red-laser line than at the violet, it can be seen that it is still quite high (~1,000,000) accounting for significant excitation and necessary compensation when used in flow cytometry.

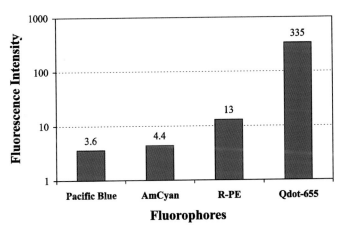

Fig. 2. Relative intensities of fluorophores with 405-nm excitation. Equimolar solutions of Qdot-655, R-PE, AmCyan, and Pacific Blue were compared in a spectrofluoro-meter (SLM 4800C) with violet excitation. The emission peak heights suggest that Qdot-655 is around 25 times brighter than R-PE, and 75–90 times brighter than Pacific Blue or AmCyan.

antibody-labeled probes, their outer layer must perform multiple functions. It must insulate the CdSe/ZnS core structure from the aqueous environment, it must prevent the nonspecific adsorption of QDs to cells, and it must provide the functional groups necessary for covalent attachment of antibodies. Recent improvements on both nanocrystal core and shell technologies have enabled production of Qdot conjugates with exceptional brightness and low nonspecific adsorption *(11)*. Quantum Dot Corp. has introduced a new surface chemistry where the polymeric shell is modified with long-chain, amino-functionalized PEGs. This new generation of Qdot nanocrystals has low nonspecific binding to cells and can be directly conjugated to antibodies through the introduced amino groups, using bifunctional cross-linkers (*see* **Subheading 2.1.**).

1.5. Multicolor Flow Cytometry

Recent publications have described flow experiments using 10 and even 17 colors within a single sample *(12,13)*. However, the spectral properties of many existing fluorophores impose certain limitations on multicolor detection. Differences in fluorophore absorption characteristics require the use of several excitation sources simultaneously. In addition, fluorophore emission spectra tend to be broad and to tail off toward the red range, resulting in overlap between colors. Therefore, the development of new fluorophores with narrowed emission peaks and large Stokes shifts promises to lead to an increased number of simultaneously detectable colors in multicolor flow applications.

1.6. Overview of Method and Applications

Basic cell-surface staining methodology involves incubation of whole blood or peripheral blood mononuclear cells with a fluorescently labeled antibody, followed by lysis of the red cells (for whole blood only), washing, and analyzing on a flow cytometer. The techniques presented are equally applicable to organic dye, fluorescent protein, and Qdot nanocrystal reagents.

The method presented can be used to surface-stain whole blood or peripheral blood mononuclear cells using an eight-color panel, developed for immune function testing involving cytokine flow cytometry (surface and intracellular staining *[14–17]*). The method can also be used to assess spectral overlap and to determine the stain index (SI) (*see* **Fig. 3** and **Subheading 3.7.**) of the CD4 Qdot-655 conjugate when used in conjunction with conventional fluorophores: CD4 AmCyan, CD4 FITC, CD4 PE, CD4 APC and CD4 APC-Cy7 (*see* **Note 1**).

Results from a multicolor experiment in which one standard violet-laser reagent, CD4 AmCyan, was substituted with the CD4 Qdot-655 conjugate are discussed (*see* **Subheading 3.8.**; **Fig. 4**; **Note 2**), as well as results from singly-stained samples (*see* **Subheadings 3.1.–3.7.**; **Table 1**; **Note 3**).

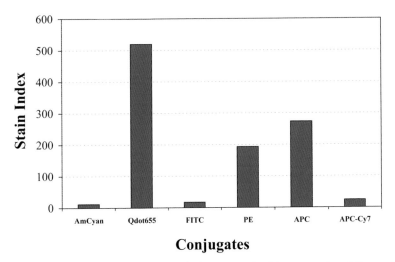

Conjugates

Fig. 3. Stain index (SI) of CD4 Qdot-655 compared with conventional fluorophores. SI is a measure of how well a signal is distinguished from background. SI takes into account the spread of the negative population and describes positive and negative population separation (*see* **Subheading 3.7.** for further explanation).

The method is applicable to general multiparameter cell-surface staining and panel development, as well as evaluation of Qdot reagents under the most stringent conditions: in combination with conventional fluorophores being analyzed on a three-laser system.

2. Materials

2.1. Conjugates

1. Monoclonal CD4 antibody (clone SK3; BD Biosciences, San Jose, CA) is custom conjugated to Quantum Dot 655 (Quantum Dot Corp.). Qdot-655 consists of CdSe/ZnS core–shell nanocrystals (ITK 2152-1) with 655-nm emission, coated with an amphiphilic carboxylated polymer and further modified with amino-functionalized PEG. Antibody attachment to the QD surface occurs through the PEG 2000-diamine spacer. The CD4 Qdot-655 conjugate to be used is prepared using an input ratio of 1:1 (1 mol of antibody per mole of Qdot nanocrystal is combined). The current technique for determining the achieved antibody-to-QD ratio is measuring depletion of antibody from the conjugation mixture in combination with spectrophotometric QD quantitation. Typically, no unconjugated antibody can be detected after labeling, suggesting that 100% is covalently coupled and that the final ratio is 1:1 (personal communication Quantum Dot Corp.).

2. Other fluorescently-labeled antibodies CD4 AmCyan, CD4 fluorescein isothiocyanate (FITC), CD4 R-phycoerythrin (PE), CD4 allophycocyanin (APC), CD4 APC-Cy7, CD57 FITC, CCR7 PE, CD28 peridinin chlorophyll-a protein

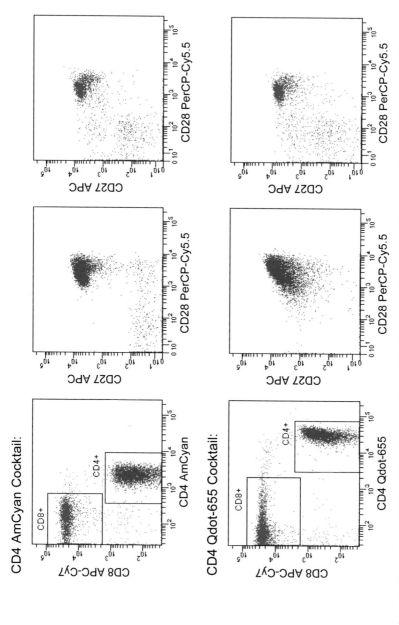

Fig. 4. Effect of CD4 Qdot-655 staining on PerCP-Cy5.5 and APC staining. Top row: CD4 Qdot-655 cocktail. Bottom row: CD4 AmCyan cocktail. Note the spreading of the positive population in PerCP-Cy5.5 and APC in the lower middle panel (CD4+ cells), and spreading of the negative populations into the *x*- and *y*-axes. The CD8+ populations (right panels), which do not bind CD4 Qdot-655, are largely unaffected. Also, the CD4+ and CD8+ populations in other channels (not shown) was largely unaffected by the use of CD4 AmCyan versus CD4 Qdot-655.

Table 1
Compensation Matrix for Typical Fluorophores in Combination With Qdot-655[a]

Detector fluorophore	FITC	PE	APC	AmCyan	Qdot-655
FITC	*	44	0	3.6	0.8
PE	0.5	*	0.1	0	1.8
APC	0.1	0	*	0	1.1
AmCyan	29	9.8	0	*	4.2
Qdot-655	0.1	0.1	96	0.1	*

[a]Detector signal—% fluorophore signal; e.g. the PE channel requires 44% of the FITC signal to be taken away from the apparent PE signal to reflect only PE fluorescence; *see* **Note 6**.

(PerCP)-Cy5.5, CD45RA PE-Cy7, CD27 APC, CD8 APC-Cy7, and CD3 Pacific Blue were from BD Biosciences (*see* **Note 1**).

2.2. Reagents

1. Whole blood, collected in EDTA tubes.
2. Erythrocyte lysis and cell-fixation buffer (e.g., BD FACS Lysing Solution™; BD Biosciences).
3. Wash buffer: PBS, 0.5% bovine serum albumin, and 0.1% NaN_3.
4. Anti-mouse Igκ and negative control compensation particles (BD™ CompBeads; BD Biosciences).
5. Fixation stabilization buffer (BD™ Stabilizing Fixative; only required for Cy7 tandems. Otherwise, 1% paraformaldehyde in phosphate-buffered saline may be substituted).

2.3. Equipment

1. 12×75-mm Polystyrene tubes (BD Discovery Labware, Bedford, MA).
2. Flow cytometer equipped with violet, blue, and red lasers and at least 10 parameter detection (2-violet fluorescence channels, 4-blue fluorescence channels, 2 red-fluorescence channels, and forward and side light scatter from the blue laser; e.g., BD™ LSR II with FACSDiVa™ Software v3.1 or later) from BD Biosciences.
3. Ultraviolet-visible spectrophotometer (Hitachi U-2001 or equivalent).
4. Spectrofluorometer, equipped with xenon-arc lamp, and photomultiplier tube (PMT) sensitive to 900 nm (SLM 4800C or equivalent).
5. Vortex mixer (e.g., Vortex-Genie 2, Scientific Industries, Bohemia, NY).

3. Methods

This chapter presents a method of cell-surface staining, utilizing directly conjugated CD4 Qdot-655 (custom conjugated by Quantum Dot Corp.) as an example of a Qdot reagent applicable to flow. The method, as well as the associated flow cytometric data acquisition and analysis, can be used with Qdot

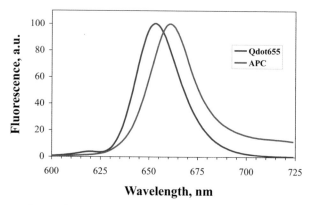

Wavelength, nm

Fig. 5. Comparison of APC- and Qdot-655-emission peak widths. Qdot-655 and APC have nearly identical peak widths at half-height (26–28 nm). A difference in spectral shape between them suggesting narrower emission on the part of QD is quantified by measuring the width at 15% of maximum peak height, where one can measure 48 nm for Qdot-655 and 65 nm for APC. If cross-talk from multiple laser excitation is not important (e.g., if Qdot reagents are used in a multicolor panel excited by a single laser), they should require less compensation.

reagents alone, conventional fluorophore reagents alone, or a combination of the two. Considerations relevant to flow cytometry in a multiple laser system where conventional fluorophores are used together with Qdot conjugates, based on experimental results obtained from the method, are also discussed.

3.1. Spectrophotometric Characterization of CD4 Qdot-655 Conjugate

1. Emission spectra. Fluorescence emission spectra, normalized to full scale, reveal similar peak widths at half-height between Qdot-655 nanocrystals and conventional fluorophores (*see* **Fig. 5**). However, the common "tailing" seen with organic dyes and fluorescent proteins is virtually absent with QDs, making them suited to multicolor flow cytometry because of less spillover between neighboring colors (for limitations because of absorption characteristics, *see* **Subheading 3.6.2.**).
2. Fluorescence intensity. Spectrophotometric analysis of CD4 Qdot-655 conjugates, CD4 Pacific Blue and CD4 AmCyan conjugates (all excited at 405 nm in a spectrofluorometer, *see* **Fig. 2**) demonstrates that Qdot nanocrystals are extremely bright fluorophores. Qdot-655 conjugate has 80–100X the fluorescence of Pacific Blue or AmCyan.

3.2. Titration of CD4 Qdot-655 Conjugate for Cell Staining

1. The target concentration of CD4 Qdot-655 reagent to use for cell staining is determined from titration, in which the Qdot concentration is varied over a range of 8.3 to 0.13 n*M* final concentration in the staining mix. Intermediate concentration points are produced by serial (twofold) dilution. An example of the titration curve

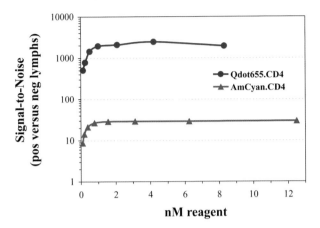

Fig. 6. Comparative titrations of AmCyan and Qdot reagents. The affinity of antibodies does not appear to be affected by conjugation with Qdots. Titration results reveal that the concentration of reagent required to achieve half-saturation is sub-nanomolar for either fluorophore. This is a typical result for CD4 AmCyan, where saturation is achieved with the equivalent of 30 ng of antibody, or less.

of CD4 Qdot-655 is shown in **Fig. 6**. Staining is according to the method described in **Subheadings 3.3.–3.5.**). For subsequent studies, a target final concentration of 5 nM is employed (*see* **Note 4**), based on the identification of where full saturation occurs from this data.

2. The surface concentration of CD4 antigen (~47,900 per CD4+ lymphocyte) was measured using QuantiBRITE PE beads *(18)*.

3. It is important to determine whether covalent attachment of antibody to a nanometer-size QD (*see* **Fig. 7**) impacts the avidity and nonspecific binding of the conjugate. A comparative titration of CD4 Qdot-655 and CD4 AmCyan reagents reveals that the concentration of reagent required to achieve half-saturation is 1–2 nM for either fluorophore (*see* **Fig. 6**) and, therefore, the affinity of antibodies does not appear to be affected by conjugation to Qdot nanocrystals.

3.3. Sample Collection

1. Collect whole blood in EDTA and store at ambient temperature up to 8 h prior to use.

3.4. Sample Processing and Staining

1. Add 100 μL of whole blood to each 12 × 75-mm tube.
2. Stain blood for 30–60 min in the dark with either CD4 FITC, CD4 PE, CD4 APC, CD4 APC-Cy7, CD4 AmCyan, and CD4 Qdot-655 as single stains, or CD4 Qdot-655 together with: CD57 FITC, CCR7 PE, CD28 PerCP-Cy5.5, CD45RA PE-Cy7, CD27 APC, CD8 APC-Cy7, and CD3 Pacific Blue as a multicolor stain. In order to test the feasibility of replacing any single reagent with the Qdot equivalent,

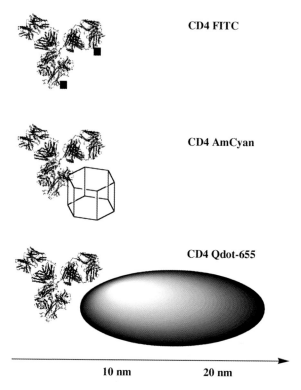

Fig. 7. Size comparison of Qdot and conventional fluorophore conjugates—FITC and AmCyan.

another tube may be stained with the same panel of eight reagents as above, but using, for example, CD4 AmCyan instead of CD4 Qdot-655. This tube should be run using the standard AmCyan mirror and filter, and a separate set of compensation reagents should be included.

3. Add 2 mL of 1X FACS lysing solution; mix at medium speed on vortex mixer. Incubate at room temperature for 10 min.
4. Centrifuge at 300–500*g* for 5 min and decant supernatant.
5. Agitate the tube to promote resuspension of cell pellet in remaining liquid. Add 2 mL of wash buffer, vortex, then centrifuge, and decant as in **step 4**.
6. Again agitate tube as in **step 5** and resuspend in 0.5 mL of wash buffer or 150 µL of fixation stabilization buffer and hold at 4°C until sample acquisition.
7. Prepare one stained sample containing BD™ CompBeads anti-mouse Igκ and negative control compensation particles from each one of the different conjugates to be used in the experiment. Treat the beads identically to the stained cell samples. This set of samples will be used to adjust or evaluate compensation. Alternatively, a set of singly-stained samples of whole blood (prepared as described above) may be used.

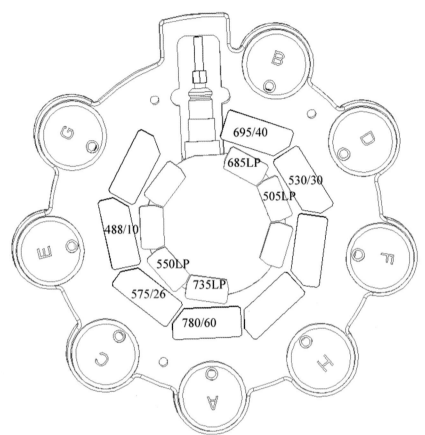

Fig. 8. Blue laser detector map showing dichroic mirror and bandpass filters used in standard configuration of BD LSRII™; detector/fluorophore assignments are as follows: A, PE-Cy7; B, PerCP-Cy5.5; C, PE; D, FITC; E, orthogonal (side) light scatter.

3.5. Data Acquisition and Analysis

1. Perform flow cytometry using a cytometer equipped with blue (488 nm), red (633 nm), and violet (405 nm) lasers (e.g., BD LSRII™).
2. In order to read Qdot-655 conjugates, configure the cytometer with appropriate dichroic mirrors and bandpass filters. For example, a standard LSRII in which either the AmCyan or the Pacific Blue PMT is equipped with a 635 LP dichroic mirror and a 655/20 bandpass filter (*see* **Note 5**; **Figs. 8** and **9**).
3. Set up the cytometer (**steps 3–8**) by first using unstained cells to set forward and side scatter PMT values. Create a one-dot plot for forward vs side scatter and create one histogram for each fluorescence parameter (e.g., on a global worksheet in BD FACSDiva™ software).
4. Set an appropriate threshold on forward scatter to eliminate debris while still retaining the lymphocyte population. Draw a gate on the lymphocyte population in a dot

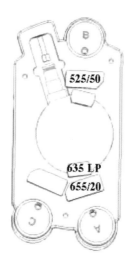

Fig. 9. Violet-laser detector map showing an example configuration applicable to analyzing Qdot-655 and AmCyan with violet-laser excitation; detector A, Qdot-655; detector B, AmCyan.

    plot of forward vs side scatter; gate the fluorescence parameter histograms on this population. Create two gates on each histogram, one for the negative population and one for the positive. Create statistics views for each histogram such that means are calculated for the gated populations in each fluorescence parameter.

5. Set the number of events to display to 2000 and the number of total events to record at 10,000.

6. Set the PMT voltages for each fluorescence parameter using unstained cells, gating on lymphocytes in light scatter, and placing the lymphocyte peak at a mean of about 100 (the peak straddles the first and second decades).

7. Test to make sure that all colors will be on scale. This can be done either by analyzing cells singly stained with the brightest fluorochrome (e.g., Qdot conjugates) and, if necessary, lowering the PMT voltage for that fluorochrome such that the positive peak mean is comfortably on scale, for example around 100,000 in linear fluorescence. Alternatively, after first setting the PMT voltages using unstained cells, a sample stained with the entire reagent panel can be run and, where necessary, the voltage of any of the PMTs can be lowered such that the brightest events are still contained within the last full decade displayed (i.e., on scale). A further alternative to **steps 6** and **7** is bead-based calibration. If used, supplier recommendations should be followed (*see* **Note 6**).

8. Adjust compensation by using whole blood samples or BD™ CompBeads stained individually with each reagent (*see* **Subheading 3.4., step 7**), and by using the AutoComp option of FACSDiVa™. Either type of sample will provide positive populations from which spectral overlaps are automatically calculated.

9. Acquire data from samples; analyze data with appropriate software. If data is analyzed with FACSDiVa™ Software v3.1 or later, it may be displayed with a

transformation that allows for improved visualization of events that appear at the lower end of the log scale (i.e., BiExponential analysis).
10. Create appropriate combination of dot plots, histograms, gates, and statistics views for the application.

3.6. Compensation

1. Compensation is a process by which spillover fluorescence is removed from secondary parameters so that fluorescence values for a parameter reflect only the fluorescence of the primary fluorophore *(19,20)*.
2. A major limitation of QDs in flow applications is the inherently broad excitation spectra (*see* **Note 7**; **Fig. 1**). It can be seen that the efficiency of excitation varies with the wavelength of the laser line (ultraviolet-strongest to red-weakest). However, the Qdot nanocrystal molar extinction coefficients are so high (e.g., six million at the violet-laser line) that even at 633 nm, where excitation is far from optimal, the extinction of Qdot-655 ($850,000\text{-}M^{-1}cm^{-1}$) is higher than that of APC, ($E_{650} = 700,000\text{-}M^{-1}cm^{-1}$). The practical consequence of these observations are that Qdot conjugates can be excited by multiple lasers and therefore require significant compensation (*see* **Table 1**; **Fig. 10**).

3.7. Separation of Positive and Negative Cell Populations; SI

1. The signal-to-noise ratio is one approach to quantifying the separation between two stained populations; it tells us how well the means (or medians) of these populations are separated. For most applications, the signal-to-noise ratio alone is insufficient to judge performance. The reason for this is that the spread of the negative, or "noise," population will also strongly impact the ability to distinguish the signal and noise populations. SI *(21,22)* measures distinguishability of populations by incorporating both the spread of the negative peak and the difference (rather than ratio) of the means (or medians) of the positive and negative populations. The negative population (also referred to as "background") may be defined according to the application; e.g., the negative peak of a stained sample, unstained cells, a cell type other than the target population in the sample, and so on).

$$StainIndex = \frac{mean_{positive} - mean_{background}}{2 \times SD_{background}}$$

2. When CD4 Qdot-655-stained cells are excited with the violet laser at 405 nm and compared with CD4 PE-stained cells excited with the blue laser at 488 nm, Qdot-655 is about 3.5 times brighter than PE on the basis of positive lymphocytes (*see* **Table 2**). An SI comparison also demonstrates the superior separation of the Qdot reagent, where CD4 Qdot-655 has an SI of 2.7 times that of CD4 PE (*see* **Fig. 3**).
3. When CD4 Qdot-655-stained cells are excited with the violet laser at 405 nm and compared with CD4 APC-stained cells excited with the red laser at 633 nm, Qdot-655 is about two times brighter than APC (*see* **Table 2**). This is reflected in data demonstrating that the SI comparison is similar, with CD4 Qdot-655 having an SI of 1.9 times higher than that of CD4 APC (*see* **Fig. 3**).

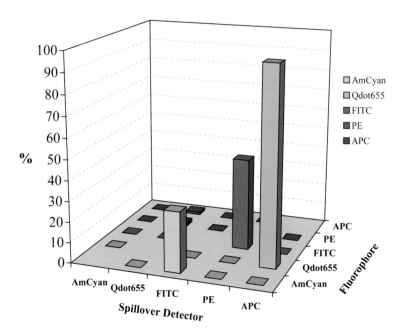

Fig. 10. Spillover of Fluorophores with Qdot-655. Qdot-655 spillover into APC is the most prominent in this group of reagents and detectors. The percentage is about 100 where other typical fluorophores may have 50% or less, such as AmCyan into FITC or FITC into PE, as shown.

Table 2
Relative Brightness of CD4 Qdot-655 Compared With Typical Fluorophores Used in Cell-Surface Staining

Fluorophore sample	FITC	Pacific blue	PE	APC	Qdot-655
Positive lymphs	1641	2729	24,861	48,477	85,729
Negative lymphs	38	45	37	26	35

3.8. Spillover Effects and SI in Multicolor Staining

1. The spillover of other fluorochromes into the Qdot-655 channel is minimal (all compensations <1.5%, mostly less than 0.1%). The spillover of the CD4 Qdot-655 into other channels, however, varies widely (*see* **Table 3**). Predictably, it has the greatest negative impact on APC (detected with a 660/20 BP filter), followed by PerCP-Cy5.5 (detected with 695/40 BP filter). **Figure 4** shows the effect of this spillover on the staining of CD28 PerCP-Cy5.5 and CD27 APC where in the CD4 positives, the CD28+CD27+ population is spread, and the CD27-populations are impossible to recognize. In general, the spillover into these channels, and the

Table 3
**Compensation Values Obtained From Setting the LSRII Flow Cytometer
for Eight-Color Staining[a]**

Channel	Compensation vs Qdot-655
FITC	0.04%
PE	0.26%
PerCP-Cy5.5	18.92%
PE-Cy7	0.51%
APC	54.79%
APC-Cy7	0.01%
Pacific Blue	0.08%

[a]The percentage reflects Qdot-655 spillover into all of the other fluorescence channels.

Table 4
**Effect on Stain Index and Signal-to-Noise When CD4 Qdot-655 is Substituted
for CD4 AmCyan in an Eight-Color Cocktail**

Stain	Stain index	Signal to noise
CD4 AmCyan in eight-color cocktail (surface stain)	7.8	23.0
CD4 Qdot in eight-color cocktail (surface stain)	145.8	786.8
CD4 Qdot alone (surface stain)	272.3	2137.8

subsequent compensation to remove it, leads to spreading of populations and, specifically, a loss of resolution among dim populations *(23)*.

2. CD4 Qdot-655 is very bright, comparable to PE in SI (this index takes into account the spread of the negative population, *see* **Table 4**). It works as part of an eight-color-staining cocktail, although spillover into PerCP-Cy5.5 and APC channels limits the ability to use these channels for high sensitivity.

4. Notes

1. APC and PE are both light-harvesting phycobiliproteins derived from *Spirulina platensis* and red algae, respectively. In algae, phycobiliproteins are incorporated into a supramolecular complex called phycobilisome, which in turn is a part of photosynthetic complex (*see* **Table 5**). APC has an absorption maximum at 650 nm, making it a good choice for red-laser (633 nm) excitation, whereas PE has absorption peaks at 496 and 565 nm making it compatible with either blue (488 nm) or green (532 nm) excitation. In both proteins, chromophores are covalently bound *(24)*. PerCP is derived from a dinoflagellate and consists of noncovalently bound peridinin and chlorophyll-a and a nonoligomeric protein *(25)*. AmCyan is one of the reef coral fluorescent proteins derived from *Anemonia majano (26,27)*. AmCyan that we used for conjugation is a recombinant protein that is composed of four subunits. The chromophore consists of three oxidized amino acid residues Met-Tyr-Gly (*see* **Table 5**).

Table 5
Characteristics of Phycobiliproteins, Fluorescent Proteins, and Tandem Fluorophores

Fluorophore	Chromophores	Subunit composition	Abs Max (nm)	Em Max (nm)	Extinction coefficient ($M^{-1}cm^{-1}$)	Molecular weight (Da)
R-phycoerythrin (PE)	Phycoerythrobilin Phycourobilin	$(\alpha\beta)_6\gamma$	496 564	578	2,000,000 at 564 nm	270,000
Allophycocyanine (APC)	Phycocyanobilin	$(\alpha\beta)_3$	650	660	700,000	110,000
Peridinin chlorophyll-a protein (PerCP)	Peridinin Chlorophyll	α_2	482	678	338,000	35,000
PE-Cy7 (tandem fluorophore)	Phycoerythrobilin Phycourobilin Cy7 dye	$(\alpha\beta)_6\gamma$	496 564	785	2,000,000 at 564 nm	270,000
APC-Cy7 (tandem fluorophore)	Phycocyanobilin Cy7 dye	$(\alpha\beta)_3$	650	785	700,000	110,000
PerCP-Cy5.5 (tandem fluorophore)	Peridinin Chlorophyll Cy5.5 dye	α_2	482	695	338,000	35,000
AmCyan (fluorescent protein)	Three oxidized amino acid residues -Met-Tyr-Gly	α_4	457	491	39,000	110,000

2. In the eight-color cocktail tested (*see* **Fig. 4**) AmCyan would still be preferred over Qdot-655 for the CD4 reagent. The benefit of Qdot-655´s increased brightness is outweighed by the exacerbation of the compensation required for APC and PerCP-Cy5.5, and the resulting loss of sensitivity in those channels.

3. Compensation values are highly dependent on relative PMT settings. Independent of these settings, it is virtually impossible to define acceptable compensation or spillover by a simple percentage. Wherever compensation is used, some sensitivity will be lost in the primary fluorescence channel. Ultimately, the user will need to determine how much loss a given application, including specifics of reagents and instrument, can tolerate *(27)*.

4. It is suggested that a working solution of Qdot–antibody conjugate be prepared for staining in PBS + 0.2% gelatin + 0.1% NaN_3 bottling buffer. However, the stability of Qdots formulated has not been investigated. The concentration of this (bottled) reagent will depend on the format employed. For example, in the method, 5 µL reagent is added to 100 µL of blood. In order to utilize this format, the Qdot-bottling concentration must be 21X the target concentration in staining. The target concentration of Qdot reagent to use for staining is determined from titration (*see* **Subheading 3.2.**). Converting the titration results from molarity with the assumption that CD4 Qdot-655 is 1 mol of antibody per mole of QD, a saturating level of reagent is around 30 ng per 100 µL blood, which is consistent with this antibody when it is conjugated with conventional fluorophores (e.g., *see* **Subheading 2.1.**).

5. Either the AmCyan or Pacific Blue detector on the violet laser needs to be equipped with a 635LP dichroic mirror and a 655/20 bandpass filter for optimal detection of Qdot-655 emission; the blue and red laser mirror and filter combinations are standard BD LSRII (*see* **Figs. 8** and **9**).

6. The instrument setup outlined in this method is common, however it is not necessarily applicable to all analyses. Because the observed fluorescence in a given detector can depend on a number of variables (e.g., the excitation and emission characteristics of the unstained cells, the laser used for excitation, the bandpass filters and dichroic mirrors used, and the sensitivity profile of the PMT), it may not be possible to achieve the same mean value in each detector as suggested in **Subheading 3.5., step 6**. Notably, autofluorescence is less with red excitation. Also, setting a detector based on the autofluorescence of cells, which can vary from donor-to-donor and day-to-day can cause the apparent brightness of a positive population to vary as well, affecting required compensation. Variation in brightness of the positives may be misinterpreted as being of biological origin or indicative of reagent instability. More standardized approaches are gaining acceptance and should make compensation percentages and brightness results more meaningful across instruments and individual experiments. Setups that optimize PMT voltages for maximum sensitivity and then routinely set PMT voltages to obtain predetermined target values of fluorescence from microparticles (e.g., BD Sphero™ Calibration Particles) may eliminate many of these problems.

7. The fluorescence excitation spectrum of Qdot-655 nanocrystals (*see* **Fig. 1**) demonstrates why most Qdot reagents, on a multilaser instrument in combination

with conventional fluorophores, will require significant compensation between detectors. Examples of this are presented in **Tables 1** and **3**, and **Fig. 10**, where the Qdot-655 most notably requires 55–96% compensation out of APC.

Acknowledgments

The authors would like to thank Dr. Holden T. Maecker for providing data on multicolor staining and protocol development and Drs. James Bishop, Alan Stall, and Vernon Maino for critical review of the manuscript.

References

1. Henglein, A. (1982) Photochemistry of colloidal cadmium sulfide. 2. Effects of adsorbed methyl viologen and of colloidal platinum. *J. Phys. Chem.* **86,** 2291–2293.
2. Brus, L. E. (1983) A simple model for the ionization potential, electron affinity, and aqueous redox potentials of small semiconductor crystallites. *J. Chem. Phys.* **79,** 5566–5571.
3. Spanhel, L., Haase, M., Weller, H., and Henglein, A. (1987) Photochemistry of colloidal semiconductors. Surface modification and stability of strong luminescing CdS particles. *J. Amer. Chem. Soc.* **109,** 5649–5662.
4. Hines, M. A. and Guyot-Sionnest, P. (1996) Synthesis and characterization of strongly luminescing ZnS-capped CdSe nanocrystals. *J. Phys. Chem.* **100,** 468–471.
5. Dabbousi, R. O., Rodriguez-Viejo, J., Mikulec, F. V., et al. (1997) (CdSe)ZnS core-shell quantum dots: synthesis and characterization of a size series of highly luminescent nanocrystallites. *J. Phys. Chem. B* **101,** 9463–9475.
6. Peng, X., Schlamp, M. C., Kadavanich, A. V., and Alivisatos, A. P. (1997) Epitaxial growth of highly luminescent CdSe/CdS core/shell nanocrystals with photostability and electronic accessibility. *J. Amer. Chem. Soc.* **119,** 7019–7029.
7. Bruchez, M., Moronne, M., Gin, P., Weiss, S., and Alivisatos, A. P. (1998) Semiconductor nanocrystals as fluorescent biological labels. *Science* **281,** 2013–2015.
8. Chan, W. C. W. and Nie, S. (1998) Quantum dots bioconjugates for ultrasensitive nonisotopic detection. *Science* **281,** 2016–2018.
9. Doose, S. (2003) Single molecule characterization of photophysical and colloidal properties of biocompatible quantum dots. Dissertation, Ruprecht-Karls University, Heidelberg, Germany.
10. Abrams, B. and Dubrovsky, T. (2005) Evaluation of the new Qdot-655 conjugates. Unpublished results.
11. Larson, D. R., Zipfel, W. R., Williams, R. M., et al. (2003) Water-soluble quantum dots for multiphoton fluorescence imaging *in vivo. Science* **300,** 1434–1436.
12. Roederer, M., DeRosa, S., Gerstein, R., et al. (1997) 8 color 10-parameter flow cytometry to elucidate complex leukocyte heterogeneity. *Cytometry* **29,** 328–339.
13. Perfetto, S. P., Chattopadhyay, P. K., and Roederer, M. (2004) Seventeen-color flow cytometry: unraveling the immune system. *Nature* **4,** 648–655.
14. Nomura, L. E., Emu, B., Hoh, R., et al. (2005) Independent loss of IL-2 production and altered differentiation of HIV-specific CD8+ T cells. *J. Immunol.* (submitted for publication).

15. Maino, V. C. and Maecker, H. T. (2004) Cytokine flow cytometry: a multiparametric approach for assessing cellular immune responses to viral antigens. *Clin. Immunol.* **110**, 222–231.
16. Maecker, H. T. (2004) Cytokine flow cytometry. In: *Methods in Molecular Biology: Flow Cytometry Protocols*, (Hawley, T. S. and Hawley, R.G., eds.), Humana, Totowa, NJ, pp. 95–107.
17. Suni, M. A., Dunn, H. S., Orr, P. L., et al. (2003) Performance of plate-based cytokine flow cytometry with automated data analysis. *BMC Immunology* **4**, 9–21.
18. Davis, K. A., Abrams, B., Iyer, S. B., Hoffman, R. A., and Bishop, J. E. (1998) Determination of CD4 antigen density on cells: role of antibody valency, avidity, clones, and conjugation. *Cytometry* **33**, 197–205.
19. Shapiro, H. M. (2003) *Practical Flow Cytometry*, 4th ed., Wiley-Liss, New York.
20. Roederer, M. (2001) Spectral compensation for flow cytometry: visualization artifacts, limitations, and caveats. *Cytometry* **45**, 194–205.
21. Bigos, M., Stovel, R., and Parks, D. (2004) Evaluating multi-color fluorescence data quality among different instruments and different laser powers – methods and results. *Cytometry* **59A**, 42.
22. Maecker, H. T., Frey, T., Nomura, L. E., and Trotter, J. (2004) Selecting fluorochrome conjugates for maximum sensitivity. *Cytometry* **62A**, 169–173.
23. Bishop, J. E., Dickerson, J., Stall, A., et al. (2004) A setup system for compensation BD CompBeads plus BD FACSDiva Software. *BD Biosciences White Paper*, San Jose, CA.
24. Glazer, A. N. (1981) Photosynthetic accessory proteins with bilin prosthetic groups. *The Biochemistry of Plants* **8**, 51–96.
25. Prézelin, B. B. and Haxo, F. T. (1976) Purification and characterization of peridinin chlorophyll-a proteins from the marine dinoflagellates *Glenodinium sp.* and *Gonyaulax polyedra*. *Planta* **128**, 133–141.
26. Matz, M. V., Fradkov, A. F., Labas, Y. A., et al. (1999) Fluorescent proteins from nonbioluminescent *Anthozoa* species. *Nat. Biotechnol.* **17**, 969–973.
27. Yanushevich, Y. G., Staroverov, D. B., Savitsky, A. P., et al. (2002) A strategy for the generation of non-aggregating mutants of *Anthozoa* fluorescent proteins. *FEBS Letters* **511**, 11–14.

V

Biochemical Applications

17

Luminescent Biocompatible Quantum Dots

A Tool for Immunosorbent Assay Design

Ellen R. Goldman, H. Tetsuo Uyeda, Andrew Hayhurst, and Hedi Mattoussi

Summary

We have developed several conjugation strategies based on noncovalent self-assembly for the attachment of proteins and other biomolecules to water-soluble luminescent colloidal semiconductor nanocrystals (quantum dots [QDs]). The resulting QD–protein conjugates were employed in designing a variety of bioinspired applications, including single and multiplexed immunosorbent assays to detect toxins and small molecule explosives. In these studies we showed that QD fluorophores offer several important advantages. In particular, their tunable broad excitation spectra combined with narrow fluorescence emission peaks permit single-line excitation of multiple color nanocrystals, with facile signal deconvolution to extract individual contributions from each population (e.g., size) of QDs in multiplexed assays. Furthermore, the QDs strong resistance to photobleaching under continuous illumination relative to many organic dyes makes them ideal fluorophores for long-term cellular imaging studies. This chapter details the materials and methods for the synthesis of surface-functionalized CdSe-ZnS core–shell QDs, the construction and preparation of recombinant proteins, the conjugation of antibodies (and antibody fragments) to QDs, and the use of antibody-conjugated QDs in fluoroimmunoassays.

Key Words: QD fluorophores; QD–antibody conjugates; fluoroimmunoassay; competition assay.

1. Introduction

Colloidal luminescent quantum dots (QDs) have unique optical and spectroscopic properties that offer a compelling alternative to traditional fluorophores in several fluorescence-based applications, including fluoroimmuno- and fluorescence resonance energy transfer-based assays *(1–3)*. The list of unique properties and advantages offered by QD fluorophores of use in bio-oriented applications based

From: *Methods in Molecular Biology, vol. 374: Quantum Dots: Applications in Biology*
Edited by: M. P. Bruchez and C. Z. Hotz © Humana Press Inc., Totowa, NJ

on fluorescence include broad, tunable absorption spectra with high molar-extinction coefficients ($\sim\times10$–100 that of organic dyes), narrow and tunable symmetric photoluminescence spectra (full width at half max $\sim$25–40 nm) that can be made to span the ultraviolet to near-infrared, and high-quantum yields in organic and buffer solutions *(1,3,4–9)*. Because QDs can be excited over a broad range of wavelengths, it is possible to choose an excitation line far from the emission peak resulting in large achievable Stokes shifts. QD fluorophores, namely those made of CdSe-ZnS core–shell materials have exceptional resistance to photo- and chemical degradation and exhibit high photobleaching thresholds *(5,7,10)*. We have recently developed several protocols to attach QDs to several types of proteins, including full antibodies and engineered antibody fragments and demonstrated their use in fluoroimmunoassays specific for the detection of toxin proteins and/or small molecule explosives.

The first conjugation strategy is based on electrostatic self-assembly between negatively charged CdSe-ZnS QDs and positively charged regions of proteins that serve to bridge the QD and antibody *(11,12)*. The nanocrystals were rendered water soluble by substituting the surface ligands with dihydrolipoic acid (DHLA) *(3)*. The surface ligand possesses a bidentate thiol moiety, which can coordinate to the ZnS surface and an opposing COOH group that permits dispersion of the capped nanocrystals in basic buffer solutions via ionization of the carboxylic acid proton.

The second conjugation strategy is also driven by self-assembly between nanocrystals and engineered proteins but employs metal-affinity coordination of histidine (His) residues (in a poly-His tag) to the ZnS surface. This strategy is based on the coordination of His molecules to immobilized metal ions such as Cu^{2+}, Ni^{2+}, Zn^{2+}, and Co^{2+} *(13–15)*. This has lead to the development of a commonly used protocol in which proteins are routinely constructed with His tails to facilitate purification over nickel gel-purification columns *(16,17)*. Our noncovalent self-assembly techniques (electrostatic and metal-affinity coordination) provide simple and reproducible methods to conjugate luminescent QDs to proteins that serve to extend and complement existing QD-labeling methods *(1,2)*.

For QD–antibody formation employing a molecular adaptor (bridging) protein, full antibodies can bind to the nanocrystal via one of two schemes. In one scheme, they can be conjugated to QDs through an engineered bridging protein consisting of the immunoglobulin G (IgG)-binding $\beta2$ domain of streptococcal protein G modified by genetic fusion with a positively charged leucine-zipper interaction domain (PG-zb). This takes advantage of the specific interactions between the PG and Fc domain of IgG and applies to many antibodies. They can also be attached using an avidin bridge, which is previously immobilized on the QD surface via electrostatic interactions; biotinylated antibodies are required to interact with the avidin bridge in this case *(12)*. In another strategy aimed at designing compact QD–IgG conjugates that avoid the use of full antibodies and bridging protein, we have engineered antibody fragments functionalized with an oligohistidine (12 His) tail at their C-terminus and immobilized them directly on DHLA-capped QDs *(18,19)*. The main interaction is driven by metal-affinity coordination of the His tail to the CdSe-ZnS surface. In the cases discussed previously, a mixed-surface conjugation is employed where a second "inert" protein (purification tool) immobilized simultaneously with the antibody on the QD surface (via self assembly) was used to facilitate separation of the desired QD–antibody product from unlabeled antibodies. A genetically engineered maltose-binding protein appended with either a charged leucine zipper (MBP-zb) or an oligohistidine tail (MBP-His) was used. This permits titration of the number of antibody or antibody fragments attached to a single QD and control over the conjugate activity. It also serves as a simple purification tool to separate antibody-coated QDs from any unbound antibody through use of affinity chromatography. **Figure 1** shows schematic representations of two mixed-surface QD–antibody conjugates, one is made of QDs coupled to antibodies using the engineered PG-zb and the other is made of QDs coupled directly to single-chain antibody fragment (scFv) that had been appended with a His tail; both conjugates employed MBP-zb as the purification tool. Protocols for preparing QDs (both CdSe and CdSe-ZnS core–shell), conjugation of QDs to antibodies and antibody fragments using these schemes, as well as the use of antibody-conjugated QDs in fluoroimunoassays for the detection protein and small-molecule targets, are described in the following sections.

2. Materials

2.1. QD Synthesis

1. 99.99% Selenium.
2. Dimethyl cadmium ($CdMe_2$).
3. Cadmium acetylacetonate.
4. 1,2-Hexadecanediol.
5. 90–95% Trioctyl phosphine (TOP).
6. Trioctyl phosphine oxide (TOPO).
7. 90% Hexadecylamine.
8. Inert gas (nitrogen or argon).
9. Glove box.
10. Schlenk line.
11. Solvents (hexane, toluene, butanol, ethanol, methanol, dimethylformamide).
12. Diethylzinc ($ZnEt_2$).
13. Hexamethydisilathiane.

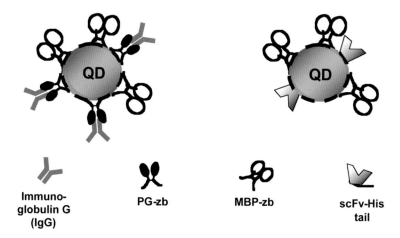

Fig. 1. Schematic representation of mixed surface quantum dot (QD)–antibody conjugates: (left) PG-zb (IgG-binding β2 domain of streptococcal protein G appended with a basic leucine zipper) acts as a molecular adaptor to connect the QD with the Fc region of IgG. (right) A mixed surface QD–antibody conjugate containing scFv appended with a poly His tail. In both cases, MBP-zb serves as a purification tool for separating QD–IgG conjugate away from excess antibody through affinity chromatography *(3,11,12)*.

14. Thioctic acid.
15. Sodium borohydride.
16. Potassium-tert-butoxide.
17. Dihydrolipoic acid (DHLA).
18. Ultra-free centrifugal filtration device, MW cut-off 50,000–100,000 (Millipore, Bedford, MA).

2.2. DNA Vector Construction and Protein Expression

1. pMal-c2 plasmid (New England Biolabs, Beverly, MA).
2. Cloning enzymes (polymerases and endonucleases).
3. QIAquick gel extraction kit (Qiagen, Valencia, CA).
4. pBad/HisB protein expression kit (Invitrogen, Carlsbad, CA).
5. *Escherichia coli* TOP 10 (Invitrogen).
6. *E. coli* Tuner (Novagen, San Diego CA).
7. Luria broth base (LB, Invitrogen).
8. Terrific broth (TB, Sigma, St. Louis, MO).
9. Ampicillin.
10. Isopropyl β-D-thiogalactoside (IPTG).
11. L-(+)Arabinose (Sigma).
12. Glucose.

2.3. Protein Purification

1. Buffer A: 100 mM NaH$_2$PO$_4$, 10 mM Tris, 6 M guanidine HCl; adjust pH to 8.0 using NaOH.
2. NiNTA resin (Qiagen).
3. 50 mL Oak Ridge polypropylene centrifuge tubes.
4. Buffer B: 100 mM NaH$_2$PO$_4$, 10 mM Tris, 8 M urea, adjust pH to 8.0 with NaOH, *immediately* prior to use.
5. Buffer C: 100 mM NaH$_2$PO$_4$, 10 mM Tris, 8 M urea, adjust pH to 6.3 with NaOH, *immediately* prior to use.
6. PBS: 200 mM NaCl, 2.7 mM KCl, 8.2 mM Na$_2$HPO$_4$, 4.2 mM NaH$_2$PO$_4$, 1.15 mM K$_2$HPO$_4$; pH 7.4.
7. Buffer E: 50 mM NaH$_2$PO$_4$, 300 mM NaCl, 250 mM imidazole, adjust pH to 6.5 with HCl.
8. Dialysis tubing (12–14 K cut off).
9. Centriprep and/or centricon (Millipore).
10. 0.22-μm Syringe filter compatible with protein samples.
11. 0.75 M Sucrose and 0.1 M Tris-HCl, pH 7.5.
12. 1 mM EDTA, pH 7.5.
13. Imidazole.
14. 0.5 M MgCl$_2$.

2.4. Immunoassays

1. Borate buffer: 10 mM sodium borate, pH 9.0.
2. Amylose-affinity resin (New England Biolabs).
3. Maltose (Sigma).
4. Small columns (such as Bio-Spin columns or Micro-Bio-Spin columns; Bio-Rad, Hercules, CA).
5. Phosphate-buffered saline (PBS) (*see* **Subheading 2.3.**).
6. 96-Well white microtiter plates (FluoroNunc™Plates MaxiSorp™surface; Nalge Nunc International, Rochester, NY).
7. Fluorescence microtiter plate reader.
8. Appropriate antibodies and antigens.

3. Methods

3.1. QD Synthesis Using Organometallic Precursors

Synthesis and growth of CdSe core nanoparticles (QDs) can be prepared using reaction of organometallic precursors at high temperatures in a coordinating solvent mixture, following the approach first developed by Murray et al. *(4,5)*. Subsequent overcoating of the core nanoparticles with a thin layer of ZnS *(6,7)* is performed to produce core–shell CdSe-ZnS QDs. Both growth and overcoating steps are described next.

3.1.1. CdSe Core

1. Prepare a 1 M stock solution of TOP:Se by dissolving 7.9 g of Se (99.99%) into 100 mL of TOP (90–95%).
2. Add 170–250 μL CdMe$_2$ and 3.5–4 mL 1 M TOP:Se to 15 mL of TOP.
3. Mix under inert atmosphere in a glove box.
4. Load into a syringe equipped with a large-gage needle for injection. Store in glove box until **step 9**.
5. Load 20–25 g of TOPO (90%) into a 100-mL three-neck round bottom flask.
6. Use a Schlenk line to heat TOPO to 150–180°C for 1–3 h under vacuum while stirring to dry and degas (*see* **Note 1**) the solvent.
7. Backfill with inert gas (typically nitrogen or argon).
8. Raise temperature to 300–340°C in preparation for precursor injection.
9. Remove flask from heating source. Retrieve the syringe from glove box and quickly inject the syringe contents into the 100-mL flask.
10. Keep the temperature below 200°C degrees for a few minutes (to avoid growth) and measure an absorption spectrum. The spectrum should show resolved features with the peak of the first transition (band-edge absorption) usually located at 470–510 nm.
11. Raise temperature to 280–300°C. These higher temperatures allow growth and annealing of the QDs.
12. During growth, periodically remove samples and measure their absorption spectra. Monitor the position of the first absorption peak and its relative width; this is indicative of a sample's size distribution. If spectra indicate that growth has stopped, raise the temperature by several degrees (if desired).
13. Once the location of the first absorption peak reaches a wavelength indicative of a desired size, lower the temperature to under 100°C to arrest crystal growth.
14. Store the growth solution in a mixture of butanol and hexane (or toluene).

3.1.2. QD Purification

To isolate QDs with TOP/TOPO-capping ligands and to obtain a sample with a more narrow size distribution, CdSe QDs are often purified using size-selective precipitation, which makes use of preferential van der Waals interactions *(4)*.

1. Retrieve a fraction of the growth solution (usually containing a mixture of QDs, TOP, TOPO, butanol, and hexane (or toluene).
2. Slowly add methanol or ethanol to precipitate the QDs.
3. Centrifugate the sample to produce a pellet of QDs and decant the excess solvent.
4. Redisperse the precipitate in hexane or toluene.
5. Reprecipitate the sample using methanol or ethanol and repeat **step 3**.

These steps should provide solutions of QDs with very low concentrations of free TOP/TOPO ligands. Repeating this operation without inducing macroscopic precipitations can substantially reduce the overall size distribution of the QDs; however, it reduces product yield *(4)*.

3.1.3. ZnS-Overcoating

In the mid-1990s a few reports *(6,7)* demonstrated that overcoating CdSe QDs with ZnS-improved quantum yields to values of 30–50%. This principle is adopted from the concept of band-gap engineering used in semiconductor physics. Coating the QDs with an additional layer of a wider band-gap semiconductor provides better passivation of surface states, and results in a dramatic enhancement of the fluorescence quantum yield *(6,7)*.

The procedure for overcoating colloidal CdSe QDs with a thin layer of ZnS can be carried out as follows: a dilute solution of QDs (containing Cd concentrations of approx 0.5 mmoles) is dispersed in a TOPO-coordinating solvent. The temperature of the solution is raised to about 150°C but kept lower than 200°C to prevent further growth of the QDs. A dilute solution of Zn (or Cd) and S precursors is then slowly introduced into the hot stirring QD solution. A typical ZnS overcoating includes the following steps:

1. Mount a round bottom flask (100 mL or larger) with a pressure-equalizing addition funnel.
2. Load 20–30 g of TOPO into the round-bottom flask and dry and degas (as described in **Subheading 3.1.1., step 6**) for 2–3 h under vacuum.
3. Add purified CdSe QD solution (dispersed in hexane or toluene) at 70–80°C to a final Cd concentration of 0.5 mmol or lower.
4. Evaporate the solvent under vacuum.
5. Increase the temperature of the QD/TOPO solution to between 140 and 180°C, depending on the initial core radius (lower temperatures are used for smaller core sizes).
6. In parallel, add equimolar amounts of diethylzinc ($ZnEt_2$) and hexamethyldisilathiane precursor that correspond to the desired overcoating layer for the appropriate CdSe nanocrystal radius to a vial containing 4–5 mL of TOP. Use inert atmosphere (e.g., glove box) to carry out this operation, as the $ZnEt_2$ is pyrophoric. (*see* **Note 2.**)
7. Load the Zn and S precursor solution from **step 6** into a syringe (in the glove box) and transfer the contents to the addition funnel.
8. Slowly add the Zn/S precursor solution to the QD/TOPO solution at a rate of about 0.5 mL/min (about 1 drop every 3–5 s).
9. Once the addition is complete, lower the solution temperature to 80°C, and leave the mixture stirring for several hours.
10. Add 10 mL of butanol and hexanes and precipitate the ZnS-overcoated QDs with methanol to recover the QD product by centrifugation.

3.2. Synthesis of CdSe-ZnS QDs Using Less Reactive Precursors

Peng and coworkers have further refined/modified the previously mentioned organometallic synthesis to make it less dependent on the presence of impurities in the TOP/TOPO and avoids the use of pyrophoric CdMe$_2$ *(20,21)*. We briefly describe this modified procedure, as this synthetic route is somewhat derived from the one developed by Murray et al. *(4)* and several of the steps involved are similar to those described previously.

1. A 100-mL three-neck round bottom flask is fitted with a septum, thermocouple adapter, and an air-cooled condenser fitted with a nitrogen inlet adapter.
2. The flask is charged with 20 g TOPO, 10 g hexadecylamine, and 5 mL TOP and the mixture is heated to 120°C under vacuum for 1–2 h.
3. The system is then stirred under nitrogen atmosphere and the temperature is raised to 360°C.
4. In a separate flask, 620 mg cadmium acetylacetonate, 1.2 g 1,2-hexadecanediol, and 10 mL TOP is heated under vacuum to 100°C until the solution becomes homogeneous.
5. The flask containing the cadmium precursor is cooled to approx 80°C and 5 mL of a 2 *M* TOP Se solution is added and mixed.
6. The cadmium and selenium solution is then rapidly injected into the hot coordinating solvent at 360°C and quickly cooled to 200°C.
7. The resulting nanoparticles can then be grown to larger sizes with additional heating to 250–280°C or can be harvested as is.
8. The solution is then cooled to 80°C and diluted with a 1:1:1 mixture of hexanes, toluene, and butanol and further cooled to room temperature.
9. The solution is centrifuged to remove any unreacted metal salts and/or impurities.
10. Overcoating of the resulting core QDs is performed as described in **Subheading 3.1.3.** with the exception that the overcoating temperature is approx 30–60°C lower for the same sized QDs based on the CdMe$_2$ route.

3.3. DHLA Cap and Water Solubilization

Water-soluble CdSe-ZnS nanoparticles, compatible with aqueous conjugation conditions, can be prepared using a stepwise procedure. A relatively thick ZnS overcoating of (more than three monolayers) should be used to prepare the water-compatible QDs.

1. Purify TOP/TOPO-capped CdSe-ZnS core–shell QDs by two to three rounds of size-selection precipitation (*see* **Subheading 3.1.2.**).
2. Suspend 100–500 mg of purified TOP/TOPO-capped QDs in 300–1000 μL freshly prepared DHLA. DHLA is prepared from thioctic acid by sodium borohydride reduction *(22)*.
3. Heat the mixture to 60–80°C for a few hours while stirring.
4. Dilute the QD solution in 3–5 mL of dimethyl formamide or methanol.
5. Deprotonate the terminal dihydrolipoic acid–COOH groups by slowly adding excess potassium-tert-butoxide. A precipitate is formed, consisting of the nanoparticles and released TOP/TOPO reagents.
6. Sediment the precipitate by centrifugation and decant the supernatant.

7. Disperse the precipitate in water. The QDs with the new DHLA caps should disperse well in the water.
8. An optional centrifugation or filtration of the dispersion (using a 0.45-μm disposable filter) permits removal of the TOP/TOPO and provides a clear dispersion of the alkyl-COOH-capped nanocrystals.
9. Use an Ultra-free centrifugal filtration devices (M_W cut-off ~50,000–100,000) to separate the DHLA-capped QDs from excess hydrolyzed K-t-butoxide and residual dimethyl formamide. This will also remove the TOP/TOPO if **step 8** is omitted.
10. Repeat the centrifugation cycle using the centrifugal filtration device four times, taking up the QD solution in water using concentration/dilution (10:1).
11. Disperse the final material in deionized water or buffer at basic pH.

Dispersions of QDs in aqueous suspension with concentrations of 5–100 μ*M* are prepared using this approach. The aqueous QD suspensions are stable for months.

3.4. MBP-zb DNA Vector Construction and Protein Expression

The coding DNA sequence for the two-domain MBP-zb was constructed using standard gene assembly and cloning techniques *(11,12)*. The detailed nucleotide coding and primary amino acid sequence of the zipper tail before engineering a poly-His tag is shown in **Fig. 2** *(11,12,23)*.

1. Amplify DNA coding for the basic zipper from the plasmid pCRIIBasic (kindly supplied by H.C. Chang of Harvard University *[24]*) using polymerase chain reaction (PCR) with the following conditions: 25 cycles (30 s at 94°C, 90 s at 60°C, and 90 s at 72°C) using primers 1 and 2 (primer 1: 5′-TGCGGTGGCTCACTCAGTTG-3′; primer 2: 5′-GCTCTAGATTAATCCCCACCTGGGCGAGTTTC-3′) and pfu DNA polymerase (Stratagene).
2. Digest-amplified DNA with *Xba*I endonuclease and ligate into the XmnI/XbaI sites within the polylinker downsteam of the *mal e* gene in the commercially available pMal-c2 vector.
3. The coding sequence for the C-terminus zipper tail (**Fig. 2**) was remodeled using standard DNA manipulation and cloning techniques to include a short spacer element linked to a hexahistidine affinity tag. The finally obtained C-terminus in pMBP-zb is shown in **Fig. 2B**.
4. The following protocol details protein expression for the MBP-zb protein.
   a. Inoculate 10 mL LB media (100 μg/mL ampicillin) with a single colony of *E. coli* (strain TOP 10; Invitrogen) freshly transformed with the pMBP-zb vector.
   b. Grow with shaking at 37°C overnight (about 15 h).
   c. Inoculate 5 mL of the overnight culture into 0.5 L LB (100 μg/mL ampicillin) and continue to grow at 37°C until an OD_{600} of about 0.5 is reached.
   d. Induce protein production by adding IPTG (from a 1 *M* sterile stock) to a final concentration of 1 m*M* and grow an additional 2 h at 37°C with shaking.

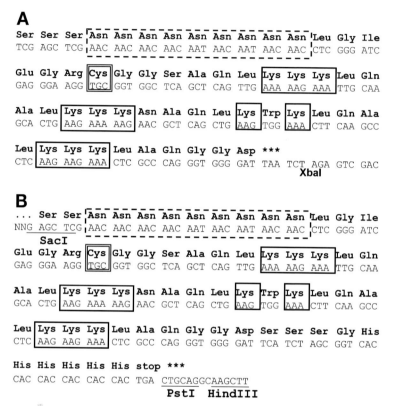

Fig. 2. (A) Primary sequence of the flexible peptide linker, and basic leucine zipper used in the construction of the MBP-zb proteins. The poly-Asn flexible linker is boxed with dashed lines, the cysteine is double boxed, and lysine residues providing net positive charge are single boxed. **(B)** Primary amino acid sequence of the appended tail used in the construction of the PG-zb proteins. It is slightly modified with respect to the one shown in **A**. The first amino acid (labeled NNG) is a Ser in the MBP-zb construction and a Glu in the PG-zb construction. The middle section is similar to what is shown in **Fig. 2A**. Restriction endonuclease cutting sites are underlined and labeled.

    e. Pellet cells by centrifugation at 2700*g* for 15 min at 4°C, and store the resulting cell pellet frozen at –80°C.

3.5. PG-zb DNA Vector Construction and Protein Expression

The two-domain PG-zb fusion protein was constructed using standard gene assembly and cloning techniques. **Figure 2B** shows the coding sequence of the zipper-containing tail used with the PG-zb. **Figure 3** shows a schematic representation of a protein-zipper construct assembled on a QD.

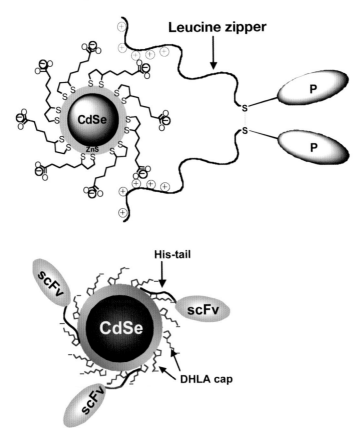

Fig. 3. (top) Schematic of a dihydrolipoic acid-capped quantum dot (QD) interacting (via electrostatic attractions) with the positively charged leucine zipper domain. (Bottom) Schematic representation of the histidine tail appended protein interacting with a ZnS-overcaoted QD via metal-histidine coordination.

1. Use PCR to amplify the β2 IgG-binding domain of streptococcal protein G (PG *[11]*) and to introduce sites for cloning with the following conditions: 25 cycles (45 s at 94°C, 45 s at 55°C, and 45 s at 72°C) using primers GNCO199 and GSAC199 (GNCO199: CAACGCTAAAATCG<u>CCATGG</u>CTTACAAACTTGT-TATTAAT; GSAC199: <u>GGTACC</u>AGATCAC<u>GAGCTC</u>TCAGTTACCGTAAAG-GTCTT, *Nco*I, *Sac*I, and *Kpn*I sites are underlined).

2. Extract *Nco*I–*Kpn*I fragment containing the PG coding sequence from a 2% agarose gel (QIAquick gel extraction kit; Qiagen) and ligate into the *Nco*I/*Kpn*I sites of expression vector pBad/HisB (Invitrogen) to produce the plasmid pBadG.

3. Ligate purified *Sac*I-*Hin*dIII DNA fragment (from pMBP-zb) which contains the coding sequences for the poly-Asn linker (from the pMal plasmid series, New

England Biolabs), a dimer-promoting cysteine, the basic leucine zipper, and C-terminal hexahistidine tag into the *Sac*I/*Hin*dIII sites of pBadG to produce pBadG-zb.

4. Screen for appropriately cloned insert.

5. The following protocol details protein expression for the PG-zb protein.

6. Inoculate 10 mL LB media (50 µg/mL ampicillin) with a single colony of *E. coli* (strain TOP 10, Invitrogen) freshly transformed with pBadG-zb.

7. Grow with shaking at 37°C overnight (about 15 h).

8. Dilute the overnight 1/100 into LB media (50 µg/mL ampicillin) and grow with shaking at 37°C until an OD_{600} of approx 0.5 is reached.

9. Induce protein production with the addition of L-(+)arabinose to a final concentration of 0.002% (w/v) and continue to grow an additional 2 h at 37°C with shaking.

10. Pellet cells by centrifugation (2700*g* for 15 min at 4°C) and store resulting cell pellet frozen at –80°C.

3.6. Construction of scFv-His Tail Construct and Protein Expression

1. Digest the original scFv-containing vector (i.e., pHen series or pCANTAB 5E) with *Sfi*I and *Not*I. Purify the insert (about 800 bp) on a 1–2% agarose gel and isolate the fragment using the QIAquick gel extraction kit.

2. Digest pMoPac45 with *Sfi*I and *Not*I (*see* **Fig. 4**).

3. Ligate the fragment containing the scFv-coding region into the pMoPac45 vector.

4. Inoculate 10 mL TB media (2% glucose, 100 µg/mL ampicillin) with a single colony of *E. coli* (Tuner, Novagen) freshly transformed with the scFv–tail plasmid and grow about 8 h at 30°C.

5. Screen for appropriately cloned insert.

6. The following protocol details protein expression for the scFv-His tail protein.
   a. Add the media grown over a day with the scFv-tail plasmid to 500 mL TB (2% glucose, 100 µg/mL ampicillin) and grow overnight at 30°C.
   b. Spin down the cells in sterile centrifuge tubes and resuspend in fresh TB without glucose (100 µg/mL ampicillin). Add IPTG to 1 mM final concentration and grow for 3 h at 25°C.
   c. Measure the final OD of the sample, and pellet the cells by spinning at 2700*g* for 15 min. Proceed immediately to the protein purification steps.

3.7. Purification of Zipper-Appended Proteins

The following protocol can be used for the protein purification of both the MBP-zb and PG-zb proteins. This is a denaturing protein preparation and serves to eliminate the copurification of nucleic acids and significant amounts of very active protease(s) that occur under nondenaturing conditions using the cytoplasmic protein fraction from cell lysis and a metal affinity column chromatography.

___Sfil(A)_____ _NcoI_ SPC _____Sfi(B)_____ __EcoRI_

G,gcc,cag,ccg,gcc,atg,gcg/tga,ggg,gcc,tcg,ggg,gcc,gaa,ttc,g

__NotI___ _SalI_ _____His6_____

cg,gcc,gca,Gtc,gac,cat,cac,cat,cac,cat,cac,gcc,ggg,tcg,gcc,

_____His6_____ ____AscI_ ____ssg3p

ggg,cat,cac,cat,cac,cat,cac,tagggcgcgccgcatagggtggtggctctgg

____ · · _____ HinDIII

ttcc. .gcgtaataaggagtcttgatatcgcaagctt. . . SKP

Fig. 4. Primary amino acid sequence of the insert region of the pMoPac45, scFv-His tail expression vector. ScFv genes or other targets are cloned as *Sfi*I/*Sfi*I, *Nco*I/*Not*I or *Sfi*I/*Not*I fragments. SPC is where the signal peptidase should cleave for expression of the mature protein. A minilinker resides between the cloning sites. The ssg3p represents an out-of-frame supershort g3p stuffer sequence used for construction purposes only. A downstream skp cistron (indicated as SKP) is included in this vector to ensure good periplasmic scFv solubility.

1. Resuspend frozen cell pellet from 500 mL *E. coli* culture in 12.5 mL denaturing buffer A and stir resuspended cells vigorously for 1 h at room temperature using a magnetic stirrer. Try to avoid excess foaming.
2. Centrifuge the suspension in a polypropylene Oak Ridge tube (50 mL) at 2700*g* for 30 min at room temperature (carrying out the above procedure at 4°C produces similar results).
3. Transfer supernatant to a clean tube.
4. Add 6 mL NiNTA resin (about 50% slurry) previously equilibrated two times with 15 mL buffer A. Tumble suspension for 45 min at room temperature.
5. Load a 1- to 1.5-cm diameter column with the protein charged resin and allow fluid to flow through. Collect and save flow-through for possible later analysis.
6. Wash settled resin in column two times with 15 mL buffer A.
7. Wash resin two times with 15 mL buffer B.

8. Wash resin four times with 15 mL buffer C.
9. Wash resin four times with 15 mL PBS buffer (pH 7.4).
10. Elute protein at about 1 mL/min with 26 mL buffer E. Collect 1-2 mL fractions. Analyze fractions by absorption A_{280} and A_{260} and/or SDS-PAGE electrophoresis. Store fractions at 4°C during analysis, and before dialysis.
11. Dialyze pooled-purified fractions against 4 L PBS at 4°C for 2–3 h, change buffer and dialyze against 4 L fresh PBS overnight.
12. Concentrate the dialyzed protein solution to about 2–3 mL using Centriprep and/or Centricon devices at 4°C.
13. Pass concentrated material through buffer-washed 0.22-μ syringe filter into a sterile plastic tube for storage. Aliquots can be stored short term at 4°C or quick frozen in a dry ice-ethanol bath for storage at –80°C.
14. Read final A_{280} on properly diluted sample. The molar extinction coefficients for the MBP-zb and PG-zb dimers are 166,200/M/cm and 35,400/M/cm, respectively. To estimate the approximate covalent monomer/dimmer ration, run a denaturing SDS-PAGE gel +/– reducing agent on final concentrated product.

3.8. Purification of scFv-His Tail Proteins

All steps must be performed using ice-cold reagents.

1. For each 20 OD of original cells, resuspend the pellet in 350 μL of 0.75 M sucrose and 0.1 M Tris-HCl, pH 7.5. If too many cells are suspended the fractionation will not be efficient (*see* **Note 3**).
2. While gently shaking the suspension add drop-wise 700 μL of 1 mM EDTA, pH 7.5 and leave rocking on ice for 15min. Avoid any local high concentrations of EDTA (*see* **Note 4**).
3. Stabilize the spheroplasts by addition of 50 μL of 0.5 M MgCl$_2$ and leave on ice for another 10 min. The MgCl$_2$ also serves to chelate the EDTA such that dialysis of the shockate prior to loading on IMAC resin becomes unnecessary.
4. Pellet the cells, and keep the osmotic shockate fraction to load on a column. Try and load the shockate immediately, because precipitation can occur within the shockate if left overnight.
5. Equilibrate an approx 5-mL Ni-NTA column with PBS.
6. Load the osmotic shockate onto the column at a rate of 1 mL/min (slower is ok). It is best to run the column in a cold room or chromatography refrigerator.
7. Wash the column with at least 200 mL PBS + 10 mM imidizole.
8. Elute with PBS + 250 mM imidizole.
9. Analyze fractions by absorption A_{280} and A_{260} and/or SDS-PAGE electrophoresis.
10. Dialyze pooled-purified fractions against 4 L PBS at 4°C for 2–3 h, change buffer and dialyze against 4 L fresh PBS overnight.
11. Centrifugally concentrate the dialyzed protein solution using Centriprep and/or Centricon devices at 4°C.
12. Run a gel on the final product and check OD_{280} and OD_{260} to determine quality and quantity of preparation. Protein can be stored at 4°C for a few months.

3.9. Immunoassays

3.9.1. Preparation of Mixed Surface QD-PG-zb/MBP-zb Conjugates

This section describes the preparation of antibody-conjugated QDs using the engineered protein PG-zb to bridge QDs and IgG antibody (*see* **Note 5**). QDs are mixed with PG-zb, the purification tool protein (MBP-zb), and IgG. The protein-coated QDs are loaded onto an amylose column used to purifiy the final QD–antibody conjugates from free antibody. **Figure 1** (left) shows a representation of the PG-zb/MBP-zb/IgG-coated QD product produced via this protocol.

1. Add 0.25 nmol of MBP-zb, 0.22 nmol PG-zb, and 0.1 nmol QDs to 200 μL borate buffer. Mix gently and incubate at room temperature for about 15 min (*see* **Notes 5 and 6**).
2. Add a second aliquot of 0.33 nmol MBP-zb to the QD–protein mix. Mix gently and incubate another 5 min at room temperature.
3. Add IgG (about 35 μg) to the QD-protein solution. Incubate at 4°C for 1 h (*see* **Note 7**).
4. Prepare an amylose column by mixing the resin and transferring 0.5 mL per column. Wash the column with at least 1 mL buffer (PBS).
5. Pipet the OD MBP-zb/PG-zb/IgG conjugates onto the top of the column. Let the QD conjugate load onto the column and immediately wash the column with 2 mL PBS.
6. Elute the MBP-zb/PG-zb/IgG-coated QDs with 1 mL of 10 m*M* maltose in PBS. The elution can be monitored with a hand-held ultraviolet light.

3.9.2. Preparation of Mixed Surface QD-scFv/MBP-zb Conjugates

1. To 200 μL borate buffer add 0.3 nmol of the scFv, 0.26 nmol of the MBP-zb purification tool protein, and 0.1 nmol of QDs. Gently mix the solution and let incubate for 15 min at room temperature.
2. Add 0.26 nmol of the MBP-zb-purification tool protein to the tube, mix gently, and incubate for another 5 min at room temperature.
3. Apply the QD-scFv preparation to a small column packed with 0.5 mL crosslinked amylose to separate any excess unbound scFv from the QD-scFv reagent and elute with a solution of 10 m*M* maltose in PBS (*see* **Subheading 3.9.1.**).

3.9.3. Protocol for Multiplexed Sandwich Immunoassays

Figure 5 shows data from a multiplexed sandwich immunoassay for cholera toxin (Calbiochem, San Diego, CA), staphylococcal enterotoxin B (SEB) (Toxin Technology, Inc, Sarasota, FL), shiga-like toxin 1(SLT) (Toxin Technology, Inc.), and ricin (Sigma). Goat anti-cholera toxin (Biogenesis, Kingston NH), monoclonal anti-SEB 6B ascites (BioVeris, Gaithersburg, MD) further purified

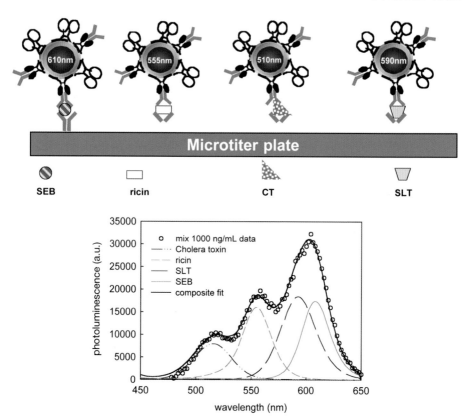

Fig. 5. Sandwich assay for the simultaneous detection of cholera toxin (CT), Staphylococal enterotoxin B (SEB), shiga like toxin (SLT) and ricin. A mix of anti-toxin antibodies were adsorbed onto wells of the 96-well plate as the capture reagent. A mix of the toxins all at equal concentrations (1000 ng/mL) was incubated with the capture antibodies. The detection reagent consisted of a mix of anti-CT antibody coupled to 510-nm emitting QDs, anti SEB antibody coupled to 610-nm emitting QDs, anti-ricin antibody coupled to 555-nm QDs and anti-SLT antibody couple to 590-nm QDs, all using the PG-zb adaptor (detailed in **Subheading 3.5.**). (Top) schematic of the assay. (Bottom) composite fluorescence signal along with the deconvoluted individual contributions of each color/toxin. (Reprinted in part from **ref. *25*** with permission from the American Chemical Society.)

by MEP Hypercel affinity chromotography (Ciphergen, Fremont, CA), a pool of monoclonal anti-SLT antibodies (9C9, 3C10, BB12) (Toxin Technology, Inc.) and monoclonal anti-Ricin RIC-03-A-G1 (a gift from the Naval Medical Research Center, Silver Springs, MD), were used as the capture antibodies adsorbed onto wells of plates *(25)*. Rabbit anti-cholera toxin antibody (Biogenesis) was coupled to 510-nm emitting QDs, polyclonal rabbit anti-ricin (a gift from the Naval Medical Research Center) was conjugated to 555-nm emitting QDs, the pool of monoclonal anti-SLT antibodies (9C9, 3C10, BB12) (Toxin Technology, Inc.) was coupled to 590-nm emitting QDs, and polyclonal rabbit anti-SEB antibody (Toxin Technology Inc.) was coupled to 610-nm emitting QDs using the PG-zb conjugation strategy detailed in **Subheading 3.9.** *(25)*.

1. Coat plates overnight at 4°C with between 2.5 and 10 μg/mL each of appropriate capture antibodies diluted into 0.1 *M* sodium bicarbonate pH 8.6 using 100 μL capture-antibody mix per well. Leave plates to incubate overnight.
2. The next day remove excess free-capture antibody and block plates with 4% (w/v) powdered nonfat milk in PBS for 1–2 h at room temperate.
3. After blocking, wash the plate twice with PBS plus 0.1% Tween-20.
4. Add 100 μL antigen solution (diluted in PBS) to wells of the plate. The antigen solution can consist of the antigen solutions alone or in mixes. It is important to include control wells with PBS only and no antigen. Test and control wells should be plated in at least triplicate.
5. Rock the plate gently at room temperature for 1 h.
6. Wash the plate twice with PBS plus 0.1% Tween-20.
7. Add 100 μL antibody-conjugated QD reagent (in PBS). Usually the QDs eluted in **Subheading 3.9.1.** are diluted to three-to-four times the elution volume.
8. Rock the plate gently at room temperature for 1 h.
9. Wash the plate two-to-four times with borate buffer (pH 9.0) containing 1% BSA. "Bang" plate to remove all liquid from the wells (*see* **Note 8**).
10. Read the fluorescence in a fluorescent plate reader. We used a Tecan Safire plate reader for data collection (Tecan US, Research Triangle Park, NC) and an excitation at 330 nm; however the broad absorption spectra of the QDs allow a wide choice of excitation lines. Choice of emission setting/filter depends on the emission spectra of the QDs used in the assay. If the control wells have a high nonspecific signal from the QDs, wells may be washed several more times with the borate buffer to further reduce nonspecific signal.
11. *See* **Note 9** for methods of signal deconvolution.

3.9.4. Protocol for Competition Immunoassay

Competition assays are frequently used in the detection of small molecules. **Figure 6** shows data from a competition immunoassay for 2,4,6-trinitrotoluene (TNT) using QDs coated with both a scFv with an engineered His tail extension and the purification tool protein (MBP-zb). Ovalbumin that had been derivatized by modification with trinitro-benzenesulfonic acid (TNB-ovalbumin) was prepared as described in **ref.** *18* and used to coat wells of a fluorescent microtiter plate. Varying amounts of soluble TNT and the anti-TNT scFv/MBP-zb-coated QDs were added to wells, incubated for an hour and then washed before quantitating the signal from the plate.

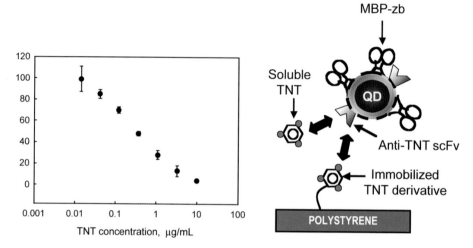

Fig. 6. (Right) Schematic description of the competition/displacement assay using a quantum dot (QD) conjugated to an anti-TNT scFv mixed with MBP-zb. An anti-TNT analog is immobilized onto the well surface. (Left) PL (in percentage) as a function of increasing free TNT added to the well along with the QD reagents.

1. Coat wells of opaque white microtiter plates overnight (4°C) with the target analog (usually conjugated to a protein) at 10 µg/mL dissolved in 0.1 M NaHCO$_3$ (pH 8.6).
2. Discard the coating solution and block the wells with 4% (w/v) nonfat powdered milk in PBS for 1–2 h at room temperature.
3. Add 50-µL portions of analyte dilutions in PBS buffer (and no analyte controls) to appropriate test and control wells.
4. Immediately add 50 µL of QD-scFv conjugate to each well. Allow the contents of the wells to come to equilibrium by incubation at room temperature with gentle shaking for about 1 h.
5. Wash the wells with borate buffer (*see* **Note 8**).
6. Measure the fluorescence with a fluorescent plate reader.

4. Notes

1. The method detailed in **Subheading 3.1.1.** describing the reaction to prepare CdSe-core nanocrystals can be scaled down to smaller volume preparations using, for example, a 50-mL flask and half of the precursors previously listed. This may be necessary when small amounts of QD materials are needed.
2. To carry out ZnS overcoating, less pyrophoric precursors such as Zn(acac)$_2$ or zinc acetate, can be used. The outcome is not very reproducible and the quality of the materials is not always as good as those produced using ZnEt$_2$.
3. The periplasmic expression of scFv at high levels is often toxic to the host cell leaving them relatively fragile. Consequently, the cells can become prone to

lysis during the osmotic-shock procedure if expression is performed at higher temperatures or for longer times. For this reason, we only express for 3 h at 25°C.

4. When we have large volumes of cells, we use an empty column that we set to slowly drip the EDTA solution into the centrifuge tubes containing the cell suspension as they gently stir on a platform shaker.

5. Mercapto-undecanoic acid and mercapto acetic acid capped CdSe-ZnS QDs have also worked with this conjugation method based on electrostatic self-assembly. Any surface ligand that leads to a negatively charged surface on the QD should also work.

6. By varying the amount of PG-zb per QD, the number of antibodies per QD can be tuned. Problems with the amylose purification, however, can arise if there are too few of the purification protein (MBP-zb) per QD.

7. If less antibody is added than there are available PG-zbs for binding antibody, generic IgG (goat IgG) can be added to the reagent before use to prevent free PG-zbs on the QD surface from binding to capture antibody in a sandwich assay.

8. These QDs contain cadmium and selenium in an inorganic crystalline form. Dispose of QD waste in compliance with applicable local, state, and federal regulations for disposal of this kind of material.

9. Measure individual spectra for each color QD used in the mix. Assay results can be deconvoluted by assuming a superposition of independent QD spectra *(25)*. A linear combination of normalized QD spectra can be used to fit the composite photoluminescence signal measured for a given assay. An n parameter-fitting function can used to determine the relative contributions of each of n QD signals to the measured composite signal (i.e., for a four-color detection):

$$I_{total}(\lambda) = \sum_{i=1}^{4} a_i I_i(\lambda) \tag{1}$$

where $I_{total}(\lambda)$ is the composite signal (corrected for background signal), a_i are fitted proportionality constants, and $I_i(\lambda)$ are the normalized individual QD spectra *(25)*. Slight adjustments (usually 2 nm or less) are made to account for minor spectral shifts of the QD bioconjugates as compared with DHLA-capped QDs. If the individual QD spectra are normalized prior to fitting, one can estimate the relative contributions of each QD signal to the composite signal. Fitted parameters are divided by their sum to give a fractional contribution estimate of each QD signal to the composite signal: $f_i = a_i/\Sigma a_i$ *(25)*.

Acknowledgments

The authors acknowledge the Office of Naval Research for the financial support, ONR grant no. N001404WX20270. We also thank A. Krishnan at DARPA for financial support. HTU was supported by the National Research Council Fellowship through NRL.

References

1. Bruchez, M., Jr., Moronne, M., Gin, P., Weiss, S., and Alivisatos, A.P. (1998) Semiconductor nanocrystals as fluorescent biological labels. *Science* **281**, 2013–2016.
2. Chan, W. C. W. and Nie, S. M. (1998) Quantum dot bioconjugates for ultrasensitive nonisotopic detection. *Science* **281**, 2016–2018.
3. Mattoussi, H., Mauro, J. M., Goldman, E. R., et al. (2000) Self-assembly of CdSe-ZnS quantum dot bioconjugates using an engineered recombinant protein. *J. Am. Chem. Soc.* **122**, 12,142–12,150.
4. Murray, C. B., Norris, D. J., and Bawendi, M. G. (1993) Synthesis and characterization of nearly monodisperse CdE (E = S, Se, Te) semiconductor nanocrystallites. *J. Am. Chem. Soc.* **115**, 8706–8715.
5. Murray, C. B., Kagan, C. K., and Bawendi, M. G. (2000) Synthesis and characterization of monodisperse nanocrystals and close-packed nanocrystal assemblies. *Annu. Rev. Mater. Sci.* **30**, 545–610.
6. Hines, M. A. and Guyot-Sionnest, P. (1996) Synthesis and characterization of strongly luminescing ZnS-capped CdSe nanocrystals. *J. Phys. Chem.* **100**, 468–471.
7. Dabbousi, B. O., Rodrigez-Viejo, J., Mikulec, F. V., et al. (1997) (CdSe)ZnS core-shell quantum dots: synthesis and characterization of a size series of highly luminescent nanocrystallites. *J. Phys. Chem. B* **101**, 9463–9475.
8. Leatherdale, C. A., Woo, W. -K., Mikulec, F. V., and Bawendi, M. G. (2002) On the absorption cross section of CdSe nanocrystal quantum dots. *J. Phys. Chem. B* **106**, 7619–7622.
9. Uyeda, H. T., Medintz, I. L., Jaiswal, J. K., Simon, S. M., and Mattoussi, H. (2005) Design of water-soluble quantum dots with novel surface ligands for biological applications. *J. Am. Chem. Soc.* **127**, 3870–3878.
10. Jaiswal, J. K., Mattoussi, H., Mauro, J. M., and Simon, S. M. (2003) Long-term multiple color imaging of live cells using quantum dots bioconjugates. *Nature Biotech.* **21**, 47–51.
11. Goldman, E. R., Anderson, G. P., Tran, P. T., Mattoussi, H., Charles, P. T., and Mauro, J. M. (2002) Conjugation of luminescent quantum dots with antibodies using an engineered adaptor protein to provide new reagents for fluoroimmunoassays. *Anal. Chem.* **74**, 841–847.
12. Goldman, E. R., Balighian, E. D., Mattoussi, H., et al. (2002) Avidin: A natural bridge for quantum dot-antibody conjugates. *J. Am. Chem. Soc.* **124**, 6378–6382.
13. Porath, J., Carlsson, J., Olsson, I., and Belfrage, G. (1975) Metal chelate affinity chromatography, a new approach to protein fractionation. *Nature* **258**, 589–599.
14. Chaga, G., Hopp, J., and Nelson, P. (1999) Immobilized metal ion affinity chromatography on Co2+-carboxymethylaspartate-agarose Superflow, as demonstrated by one-step purification of lactate dehydrogenase from chicken breast muscle. *Biotechnol. Appl. Biochem.* **29**, 19–24.
15. Hainfeld, J. F., Liu, W., Halsey, C. M. R., Freimuth, P., and Powell, R. D. (1999) Ni-NTA-gold clusters target His-tagged proteins. *J. Struct. Biol.* **127**, 185–198.
16. Hochuli, E., Dobeli, H., and Schacher, A. (1987) New metal chelate adsorbent selective for protins and peptides containing neighboring histidine-residues. *J. Chromatogr.* **411**, 177–184.

17. Hochuli, E., Bannwarth, W., Dobeli, W., Gentz, R., and Stueber, D. (1988) Genetic approach to facilitate purification of recombinant proteins with a novel metal chelate adsorbent. *Bio/Technology* **6**, 1321–1325.
18. Goldman, E. R., Medintz, I. L., Hayhurst, A., et al. (2005) Self-assembled luminescent CdSe-ZnS quantum dot bioconjugates prepared using engineered poly-histidine terminated proteins. *Analytica Chimica Acta.* **534**, 53–67.
19. Goldman, E. R., Medintz, I. L., Whitley, J. L., et al. (2005) A hybrid quantum dot-antibody fragment fluorescence resonance energy transfer-based tnt sensor. *J. Am. Chem. Soc.* **127**, 6744–6751.
20. Peng, Z. A. and Peng, X. G. (2001) Formation of high-quality CdTe, CdSe, and CdS nanocrystals using CdO as precursor. *J. Am. Chem. Soc.* **123**, 183–184.
21. Qu, L., Peng, Z. A., and Peng, X. G. (2001) Alternative routes toward high quality CdSe nanocrystals. *Nano Lett.* **1**, 333–337.
22. Gunsalus, I. C., Barton, L. S., and Gruber, W. J. (1956) Biosynthesis and structure of lipoic acid derivatives. *J. Am. Chem. Soc.* **78**, 1763–1768.
23. O'Shea, E. K., Lumb, K. J., and Kim, P. S. (1993) Peptide velcro-design of a heterodimeric coiled-coil. *Curr. Biol.* **3**, 658–667.
24. Chang, H. C., Bao, Z. Z., Yao, Y., et al. (1994) A general-method for facilitating heterodimeric pairing between 2 proteins- applications to expression of alpha-T-cell and beta-T-cell receptor extracellular segments. *Proc. Natl. Acad. Sci. USA* **91**, 11,408–11,412.
25. Goldman, E. R., Clapp, A. R., Anderson, G. P, et al. (2004) Multiplexed toxin analysis using four colors of quantum dot fluororeagents. *Anal. Chem.* **76**, 684–688.

18

Fluorescence-Based Analysis of Cellular Protein Lysate Arrays Using Quantum Dots

David H. Geho, J. Keith Killian, Animesh Nandi, Johanne Pastor, Prem Gurnani, and Kevin P. Rosenblatt

Summary

Reverse-phase protein microarrays (RPPMAs) enable heterogeneous mixtures of proteins from cellular extracts to be directly spotted onto a substrate (such as a protein biochip) in minute volumes (nanoliter-to-picoliter volumes). The protein spots can then be probed with primary antibodies to detect important posttranslational modifications such as phosphorylations that are important for protein activation and the regulation of cellular signaling. Previously, we relied on chromogenic signals for detection. However, quantum dots (QDs) represent a more versatile detection system because the signals can be time averaged and the narrow-emission spectra enable multiple protein targets to be quantified within the same spot. We found that commercially available pegylated, streptavidin-conjugated QDs are effective detection agents, with low-background binding to heterogeneous protein mixtures. This type of test, the RPPMAs, is at the forefront of an exciting, clinically-oriented discipline that is emerging, namely tissue or clinical proteomics.

Key Words: Quantum dots; protein microarray; nanotechnology; hyperspectral imaging.

1. Introduction

Reverse phase protein microarrays (RPPMAs) were developed because of the need to interrogate the state of signal pathway checkpoints in vivo from microscopic biopsy specimens *(1)*. The state of such pathways may signify misguided cellular processes, such as incipient malignant transformation. The histopathological expression of malignant transformation consists of invasion of malignant cells into the surrounding stroma, with resultant complex tissue interactions between malignant cells, blood vessels, matrix, and stroma *(2)*. The summation of these interactions is reflected in the activation of signal transduction networks

From: *Methods in Molecular Biology, vol. 374: Quantum Dots: Applications in Biology*
Edited by: M. P. Bruchez and C. Z. Hotz © Humana Press Inc., Totowa, NJ

in lesional and perilesional component cells. But, the activation (e.g., phosphorylation) status of signal pathway checkpoints is not ascertainable by gene expression profiling alone, and immunohistochemistry lacks precision in the analysis of subtle quantitative changes in multiple classes of signaling molecules that may synergize to produce a malignant phenotype. Many of the most interesting regulatory molecules will be of low abundance in the cell or serum; the difficulty in detecting low-abundant proteins partly arises from our inability to amplify the proteome, in contrast with the PCR amplification of DNA and mRNA. RPPMAs were born from the need for a highly quantitative and precise protein array technology that could debrief the dynamic protein circuitry regulating cell growth, survival, and neighboring interactions present within human tissues.

In contrast to previous protein array formats wherein a capturing probe is immobilized, RPPMAs immobilize the entire protein content of a cell population procured from the lesion of interest *(3)*. The analyte-coated array surface can then be probed with antibodies specific for individual candidate proteins of signaling pathways or other targets of interest. A typical experiment begins with laser capture microdissection (LCM) of lesional cells of interest, lysis in a suitable buffer, and spotting a few nanoliters of the protein extract onto a substrate, such as nitrocellulose-coated glass slides in defined locations with a pin-or quill-based microarrayer. Currently, up to 1000 individual cellular lysates can be arrayed on a single glass slide with a spot diameter of 250–350 μm. Each fashioned slide can then be probed with an antibody and the binding detected typically by fluorescent or colorimetric assays. The precision, specificity, and dynamic range of RPPMAs made from protein extracts on nitrocellulose slides are quantifiable and reproducible.

Most methodologies widely in use today for protein detection are limited in terms of their sensitivity, dynamic range, durability, speed, safety, and their utility for multiplexing. Presently, many RPPMAs systems rely primarily on chromogenic reporter technologies. In order for sufficient signal to be generated, these chromogenic systems require an amplification step, such as biotinyl tyramide deposition *(3)*. This limitation of the colorimetric system allows only one reporter output (such as amount of chromogen deposition) per protein spot. Fluorescent reporter molecules enjoy a relatively broad dynamic range suitable for array assays; however, because organic fluorophores are not very robust and exhibit significant bleaching with prolonged exposure to excitation, the arrays cannot be time averaged for extended periods of time. Additionally, the broad emission and narrow excitation spectra of most fluorescent reporters limits their usefulness in multiplexing assays and requires multiple excitation sources in imaging instruments, which raises costs and complicates assay design. Quantum dots (QDs) are an attractive alternative

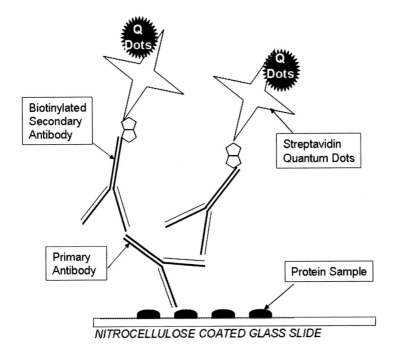

Fig. 1. Schematic outline of the steps involved in constructing a reverse-phase microarray. In the RPPMAs, proteins extracted from cellular lysates are arrayed onto a nitrocellulose substrate and probed with a primary antibody. A biotinylated secondary antibody recognizes the presence of the primary antibody. The biotinyl groups are then detected by streptavidin linked to reporter molecules, such as quantum dots instead of an enzyme such as horseradish peroxidase.

detection system for RPPMAs because they resist photobleaching while having wide excitation and narrow-emission spectra *(4)*. These characteristics hold promise for the development of a high-throughput, multiplexed RPPMA system. In the following pages, a generic methodology for the construction of a QD-dependent RPPMAs is described.

2. Materials

2.1. Equipment

1. GMS 417 pin-and-ring arrayer (Affymetrix, Santa Clara, CA) or SpotArray™ 24 Microarray Printing System (Perkin Elmer, Boston, MA).
2. FluorChem™ 9900 cabinet system (Alpha Innotech, San Leandro, CA).
3. 655-nm Narrow bandwidth emission filter (Omega® Optical, Brattleboro, VT).
4. DAKO Autostainer (Dako Cytomation, Carpinteria, CA) or ProteinArray Workstation™ (Perkin Elmer).

2.2. Supplies

1. Nitrocellulose-coated FAST® glass slides (Whatman, Sanford, ME previously Schleicher & Schuell Bioscience, Keene, NH).
2. 2X Tris-glycine SDS sample buffer (Invitrogen, cat. no. LC2676).
3. T-PER™ tissue protein extraction reagent (Pierce, cat. no. 78510).
4. β-Mercaptoethanol (Sigma, St. Louis, MO, cat. no. M-6250).
5. Protease inhibitors.
6. Phosphatase inhibitors.
7. Reblot (Chemicon, Temecula, CA).
8. Block (Applied Biosystems, Foster City, CA).
9. CSA Buffer (Dako Cytomation).
10. QDot® 655-PEG-streptavidin or non-pegylated- QDot 655-streptavidin (Quantum Dot Corporation, Hayward, CA).

3. Methods

RPPMA construction consists of four basic steps (**Figs. 1** and **2**):

1. Extraction of proteins from cells that have been procured either by laser capture microdissection or from cell cultures.
2. Application of the extracted proteins upon the solid support of the arrays.
3. Immunostaining.
4. Imaging.

The tissue source of the protein determines the protocol for lysis and sample preparation. Spotting of proteins onto the arrays, immunostaining, and imaging can be done in a completely automated fashion.

3.1. Protein Extraction From Microdissected Tissues

The extraction buffer for liberating proteins from cells that have been procured by laser capture microdissection consists of a detergent, a denaturing agent, and a buffer (*see also* **Note 1**).

1. To prepare the protein extraction buffer, prepare a 5% solution of 2-mercapto-ethanol (BME) using the 2X Tris-glycine SDS sample buffer, yielding BME-SDS buffer. Example: add 50 μL of BME to 950 μL of 2X Tris-glycine SDS sample buffer.
2. Add equal amounts of BME-SDS buffer and T-PER in order to make the extraction buffer. Example: add 1 mL of BME-SDS buffer solution to 1 mL of TPER.
3. If the proteins of interest are phosphorylated, phosphatase inhibitors may be added to the extraction buffer solution (*see* **Note 2**).
4. Pipet the desired quantity of the final extraction buffer onto the bottom of a 500-μL Eppendorf tube. The minimum volume of extraction buffer required to cover the

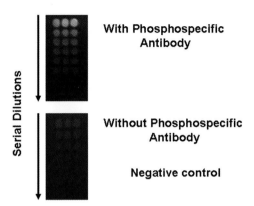

With Phosphospecific Antibody

Without Phosphospecific Antibody

Negative control

Fig. 2. Example of a reverse-phase microarray using quantum dot (QD) reporter technology. Protein extracts from a cultured lymphocyte-derived cell line were harvested and arrayed on FAST® nitrocellulose backed glass slides. The protein extracts were arrayed upon nitrocellulose slides in serial dilutions. The array surfaces were then exposed to an antibody directed toward a phosphoryated (activated) isoform of a signaling protein and then subjected to biotinyl tyramide amplification. Then QDot® 655-streptavidin reporter molecules (20 n*M*) were incubated on the arrays. For visualization, the slides probed with QDs were excited using 254-nm ultraviolet light for 2 min and were visualized using a 655-nm narrow-bandwidth (20 nm) emission filter.

surface of an LCM cap is 15 µL. The minimum number of cells for protein microarrays is approx 10,000 cells/15 µL extraction buffer.

5. Thaw each LCM cap at room temperature and remove all traces of condensation from the edges and rim of the cap with a Kimwipe tissue.
6. Place the CapSure transfer film cap containing cells captured by LCM securely onto the tube. Invert and, using a slinging motion, deposit the buffer onto the surface of the cap. Check for leaks. Mix well. *Do not vortex.*
7. Place the inverted tube with the extraction buffer resting on the surface of the cap into a 75°C ± 2 oven for 30 min to 2 h.
8. After a 15-min incubation at 75°C, gently mix the cap so the extraction buffer moves over the cap surface. Place cap back in the oven.
9. At the end of the incubation, mix the inverted tube again.
10. Place the tube into a microcentrifuge and spin the sample for approx 1 min.
11. Remove supernatant to a clean, labeled 1.5-mL screw cap tube. Discard the LCM cap.
12. Store the lysate at –20°C until time to run the downstream assay.

3.2. Protein Extraction from Cell Culture Lysates

Proteins may be similarly extracted from cells harvested under varying culture conditions using the extraction buffer described previously.

1. To prepare the protein extraction buffer, prepare a 5% solution of BME using the 2X Tris-glycine SDS sample buffer, yielding a BME-SDS buffer. Example: add 50 µL BME to 950 µL of 2X Tris-glycine SDS sample buffer.
2. Add equal amounts of BME-SDS buffer and T-PER in order to make the extraction buffer. Example: add 1 mL of BME-SDS buffer solution to 1 mL of TPER.
3. If the proteins of interest are phosphorylated, phosphatase inhibitors may be added to the extraction buffer solution (*see* **Note 2**).
4. Cells from culture are harvested and centrifuged at 1200 rpm for 10 min at 4°C. The supernatant is removed and the cells are washed several times in phosphate-buffered saline (PBS) and centrifuged at 1200 rpm for 10 min at 4°C each time.
5. The cell pellet is then resuspended in the extraction buffer and vortexed followed by incubation on ice for 20 min.
6. The lysate is then centrifuged at 10,000 rpm for 4 min at 4°C.
7. The supernatant (the protein extract) is then stored for future use at –80°C.

3.3. Spotting of Reverse-Phase Microarrays

1. Protein extracts can be arrayed on nitrocellulose-coated FAST glass slides using a GMS 417 pin-and-ring arrayer or a SpotArray 24 Microarray Printing System.
2. The lysates are arrayed in serial diltions in duplicate or triplicate. As a background control, protein extraction buffer alone is arrayed as well.
3. Approximately 1.5 nL per spot is arrayed.
4. Spatial densities of 980 spots/slide and greater can easily be accommodated on a 20–30-mm slide.
5. The slides may be stored at –20°C in a dessicated environment (Dreirite, W.A. Hammond, Xenia, OH) or used immediately.

3.4. Immunostaining Procedure

1. Prior to staining, incubate the array with 1X Reblot for 15 min.
2. Wash the arrays twice for 5–10 min with Ca^{2+}- and Mg^{2+}-free PBS.
3. Place the slides in I-Block for at least 2 h. I-Block powder is dissolved in Ca^{2+}- and Mg^{2+}-free PBS/0.1% Tween-20.
4. Slides can be immunostained using an automatic slide stainer such as the DAKO Autostainer or the ProteinArray Workstation™ using manufacturer supplied reagents.
5. Incubate the arrays for 5 min with hydrogen peroxide and then rinse with high-salt Tris-buffered saline (CSA Buffer, DAKO) supplemented with 0.1% Tween-20.
6. Block the slides with avidin block solution for 10 min, rinse with CSA buffer, and then incubate with biotin block solution for 10 min.
7. Wash with CSA buffer again and incubate for 5 min with protein block solution. Air-dry the slides.
8. The arrays are then incubated with either a specific primary antibody diluted in DAKO antibody diluent or, as a control, only DAKO antibody diluent (time must be optimized individually for each application).
9. Wash the array with CSA buffer. Incubate the array with a secondary biotinylated antibody (time must be optimized individually for each application).

10. To amplify the signal, arrays are washed with CSA buffer and incubated with streptavidin-biotin complex (DAKO) for 15 min. After another wash with CSA buffer, add the amplification reagent and incubate for 15 min followed by another wash with CSA buffer.

11. Develop the arrays using streptavidin-conjugated QDs diluted in 2% albumin in PBS. Incubation times are experimentally determined.

12. The slides probed with QDs can be visualized using 254-nm epi-fluorescent ultraviolet lamps in a FluorChem™ 9900 cabinet system. The emission signal is visualized using a 655-nm narrow-bandwidth emission filter (Omega Optical) (*see* **Note 3**).

13. A noncommercial hyperspectral imaging microscope can also be employed for detecting proteins labeled with QDs without the use of catalyzed reporter amplification (Heubschman, 2002). This technology permits multiplexing of numerous proteins in parallel (e.g., activated vs inactivated forms) without the need for specially designed filters.

    a. Arrayed slides are pretreated with 1X Re-Blot and I-Block as described in **steps 1–3** followed by avidin and biotin blocking and then incubation with the DAKO protein block solution.

    b. After a 30-min incubation with the primary antibodies, and subsequent incubation with the biotinylated secondary antibody, the slides are exposed directly to the pegylated-QDot 655-PEG-Sav reagent for 15 min (*see* **Note 4**).

    c. After washing, the slides are dried and analyzed on the hyperspectral imaging system using a spectrophotometer coupled to a charge-coupled device camera. The slit width of the spectrophotometer is 50 microns across and a 100 W Hg lamp is used for excitation.

    d. Continuous spectra between 549 and 666 nm are collected for each array, but this range can be expanded depending on the wavelength of the QDs used in the assays (*see* **Note 5**).

3.5. Conclusion

We described protocols for the use of streptavidin-conjugated QDs in RPPMAs. Via bioconjugate techniques, QDs, once bound to their targets within a tissue or analyte, effectively become nanocrystal passive integrated transponder (pit) "tags." QDs are readily addressable through facile chemical conjugation of functional moieties to the nanocrystals. Common to both QDs and older fluorescence-based probes, a fluorescence source interrogator is used to activate the passive tag with fluorescent energy and extract information from the tag. QDs overcome many performance limitations of traditional fluorescent tags. Their unique physicochemical properties address a number of issues with standard fluorescent tags, including:

1. Robustness in signal output during excitation.
2. Tendency of tissues and liquids to quench fluorescent signals.

3. Limited multiplexing capabilities secondary to broadly overlapping emission bandwidths of different tags.

QDs can withstand the physical stress of heat, light, X-rays, or irradiation that destroy older generation fluorophores. Their unique stability presents an advantage over thermo- and light-sensitive tags. Also, they have a superior range of detection and a large capture field that are required for in vivo applications.

Surpassing traditional fluorophores, QDs can be configured as memory modules with programmable read/write capabilities, and could be (re)programmed remotely, through magnetic or radio control. They are an open (detectable) system that is closable, meaning their information can be hidden or masked for patient privacy in clinical assays. Tags can be shipped from the factory in a read-only format, or that can permit field programming (through conjugation chemistry for example) by the user. QDs are "in-use" programmable because they can be manipulated and interrogated while attached to their target analytes. They provide a vast number of identification codes, again through unique and facile conjugation chemistry. All of these attributes, combined with their strong signal intensities, make QDs an exciting new reporter technology for high-throughput clinical proteomic applications such as the RPPMAs.

4. Notes

1. It is not necessary to extract proteins immediately after microdissection. The caps of microdissected cells can be stored in a clean Eppendorf tube at –80°C until all samples have been collected.
2. If the proteins of interest are particularly subject to degradation, protease inhibitors can be added to the T-PER prior to making the solution with the sample buffer. Dissolve 1 Complete™ mini protease inhibitor cocktail tablet (Roche, cat. no. 1836153) in 10 mL of T-PER. Alternatively, add 1 mL deionized water to one vial of Calbiochem protease inhibitor cocktail set I. Dilute this 100X solution to 1X (10 µL protease inhibitor cocktail for each 1000 µL of extraction buffer) for use in the extraction buffer. Then proceed with the preparation of the extraction buffer.
3. The QD-probed array was scanned using an Alpha Innotech FluorChem 9900 imager using an excitation wavelength of 254 nm and emission measured at 655 nm. Among the several scanning modalities employed, the most effective signal to noise ratio was achieved using an Omega Optical filter made for assessment of 655-nm fluorescent emissions (20-nm bandwidth centered at 655 nm with added heavy blocking of wavelengths in the infrared and ultraviolet ranges).
4. To decrease nonspecific binding, a modified form of streptavidin-conjugated QDs, QDot 655-PEG-Sav, was used as the detection molecule in the RPPMAs format. This modification involved the introduction of polyethylene glycol groups onto the streptavidin–QD conjugates. In contrast to the nonpegylated form of the

streptavidin–QD conjugate, QDot 655-PEG-Sav markedly improved the assay by decreasing the intrinsic binding characteristics of the streptavidin QDs to the protein spots.

5. The stability of QDs, in addition to their resistance to photobleaching, allows them to be imaged multiple times, for extended periods of time, in order to capture the linear portions of markedly different dilution curves located on the same array surface.

References

1. Paweletz, C. P., Charboneau, L., Bichsel, V. E., et al. (2001) Reverse phase protein microarrays which capture disease progression show activation of pro-survival pathways at the cancer invasion front. *Oncogene* **20,** 1981–1989.
2. Liotta, L. A. and Kohn, E. C. (2001) The microenvironment of the tumour-host interface. *Nature* **411,** 375–379.
3. Liotta, L. A., Espina, V., Mehta, A. I., et al. (2003) Protein microarrays: meeting analytical challenges for clinical applications. *Cancer Cell* **3,** 317–375.
4. Geho, D., Lahar, N., Gurnani, P., et al. (2005) Pegylated, steptavidin-conjugated quantum dots are effective detection elements for reverse phase protein microarrays. bioconjugate chemistry. submitted.

19

Application of Quantum Dots to Multicolor Microarray Experiments
Four-Color Genotyping

George Karlin-Neumann, Marina Sedova, Mat Falkowski, Zhiyong Wang, Steven Lin, and Maneesh Jain

Summary

Highly multiplexed genomics assays are challenged by the need for a sufficient signal-to-noise ratio for each marker scored on a microarray-detection platform. Typically, as the number of markers scored (or target complexity) increases, either more assay-target material must be applied to the array or the specific activity of each marker must be proportionately increased. However, hybridization of excessive amounts of target to the microarray can result in elevated nonspecific binding and consequent degradation of information. We have found that quantum dots provide a successful alternative to organic dyes for achieving highly multiplexed (>20,000-plex) and highly accurate, four-color genotyping and have the additional advantage of being excitable by a single wavelength of light despite their distinct emission wavelengths.

Key Words: Genotyping; single-nucleotide polymorphism; SNP; microarray; quantum dot; multiplex; targeted; molecular inversion probes; MIP; tag arrays; multicolor; unamplified genomic DNA; barcodes; padlocks.

1. Introduction

DNA microarrays provide a powerful detection medium for the simultaneous readout of many thousands of genetic markers or target sequences in a biological sample. These may represent queries of single-nucleotide polymorphisms (SNPs) or copy number at specific chromosomal locations *(1–6)*, transcript abundance among cellular RNAs *(7,8)*, DNA (or RNA) binding factor sites *(9–12)*, methylation status of specific sites in DNA *(13,14)*, or even the profiling of proteins *(15,16)*.

From: *Methods in Molecular Biology, vol. 374: Quantum Dots: Applications in Biology*
Edited by: M. P. Bruchez and C. Z. Hotz © Humana Press Inc., Totowa, NJ

We have previously developed a highly multiplexed, four-color genotyping assay using circularizing molecular-inversion probes (MIP) to interrogate un-amplified genomic DNA for more than 10,000 SNPs in a single reaction (**Fig. 1A** *[17]*). Each probe species queries a single SNP locus and is covalently bound to a unique, highly specific, synthetic tag sequence that is ultimately detected on a GeneChip® Universal Tag array in order to score its associated SNP. After anneal-ing to a genomic DNA sample, each probe is covalently circularized in a gap-fill/ligation reaction only when the proper base ("A," "C," "G," or "T") is present that complements the SNP on the genomic DNA template (*see* **Fig. 1B**). A panel of MIP probes is simultaneously reacted in four parallel reactions—one containing only the "A" nucleotide, one containing only the "G" nucleotide, and so forth, to determine which polymorphisms are present in the sample at each locus. Running all probes in each of the four reactions allows determination of both signal and assay background values, and hence a quality score (assay signal -to-background or S/B) for each SNP, and even allows detection of occasional un anticipated tri-allelic SNPs. Following amplification with channel-specific universal primers, the four reactions are combined, hybridized to a single microarray and after four-color staining and scanning, each SNP is scored according to the intensities of the fluorescent wavelengths present at its position on the array (**Figs. 1A** and **2**).

This earlier version of the assay utilized four organic dyes and a scanner with a broad-spectrum white light source to produce the fluorescent microarray sig-nals from which genotypes were derived. Each of these dyes required different excitation wavelengths from one another and therefore either multiple excitation sources or a tunable broadband source. In order to improve upon the signal-to-noise (S/N) characteristics of the assay and to enable detection on scanners with only a single excitation wavelength, we have developed a four-color microarray detection system using four different semiconductor nanocrystals (quantum dots [QDs]), each with a different emission wavelength and conjugated to a different binding protein that recognizes its specific ligand on the hybridized target. The use and performance of this system on Affymetrix GeneChip Universal Tag arrays and the GeneChip Scanner 3000 7G-4C (GCS3000-4C) are described.

2. Materials

The following materials are commercially available from Affymetrix to enable a researcher to perform 10,000 cSNP genotyping in their lab.

1. Affymetrix GeneChip Human Panel 1 10K cSNP kit: kit v1.5. (cat. no. 900866) con-tains enough reagents to process a total of 24 assays (including one control).
2. Affymetrix GeneChip Universal 10K Tag array: arrays have approx 10,000 features on each array that can detect 10,000 SNPs using the Affymetrix GeneChip DNA Analysis System incorporating MIP technology (6 pack, cat. no. 900604), (96 pack, cat. no. 900580).

A

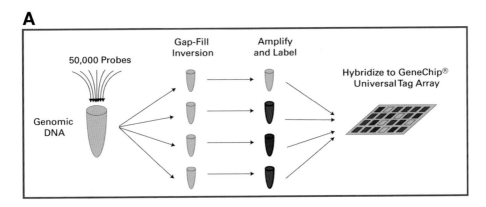

B

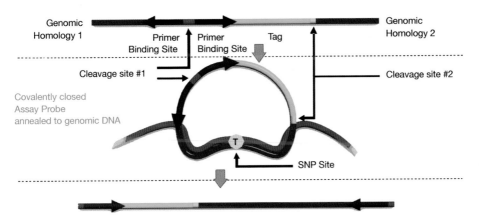

Fig. 1. Schematic of highly multiplexed, four-color molecular inversion probe (MIP)-targeted genotyping assay. **(A)** Assay process from annealing up to 50,000-plex probe pool to sample DNA, through hybridization to GeneChip Universal Tag arrays. After hybridization and washing, genotyping signals are generated by staining with quantum dot conjugates and scanning with a GCS3000 four-color scanner. **(B)** MIP before and after inversion. (Top): design of uninverted MIP probe showing genomic homology regions that will sandwich the interrogated single-nucleotide polymorphism (SNP) site, the universal PCR primer sites, the unique tag associated with the SNP marker, and several cleavage sites. (Middle): MIP probe "padlocked" onto a genomic DNA template after annealing and gap-fill/ligation at a "T" SNP site. The circularized probe is subsequently inverted by cleavage at site no. 1. (Bottom): the inverted probe is ready for PCR amplification and labeling (with universal primers), and subsequent hybridization to a universal tag array.

Quantum Dot dyes   **Organic dyes**

Fig. 2. Single excitation wavelength, GCS3000-4C confocal laser scanner effectively excites and detects multiple quantum dots with distinct emission wavelengths in a single sample. Shown are representative images of four-color, 5000-plex molecular inversion probe genotyping on Affymetrix GeneChip Universal 5K Tag arrays stained with either Qdot or organic dye conjugates (the latter imaged on a charge-coupled device scanner). The four channels of information from four sequential images are superimposed in false color. Each feature represents the genotype for a single marker. Note that different samples were genotyped on the two chips shown.

3. Affymetrix GeneChip Scanner 3000 Targeted Genotyping System: GCS 3000 TG System (cat. no. 00-0185).
4. HapMap CEPH plate DNA samples: the DNA samples were obtained from the Coriell collection (Coriell Institute, Camden, NJ) and consisted of trios from the CEU family collection. A full list of these samples is available at the HapMap website (www.hapmap.org).

3. Methods

After identifying candidate genes or regions of interest using whole-genome association or linkage studies, scientists use GeneChip custom genotyping solutions to target genes associated with disease or variable drug response. In addition, researchers can also design custom SNP assays for agricultural and other nonhuman applications.

Affymetrix offers a broad menu of products for candidate gene analysis and fine mapping, including GeneChip Universal Tag arrays, targeted genotyping reagents, and a wide range of services.

Further information about methods for genotyping based on the MIP technology are available at the Affymetrix website:http://www.affymetrix.com/products/application/targeted_genotyping.affx.

4. Results

4.1. Technical Motivation for Work

As successively larger numbers of SNP markers have been combined in a single MIP-genotyping reaction over the past 4 yr *(17)*, the development of higher specific activity targets has been a key area of investigation for us. At the same time, we wished to enable the four-color MIP assay to be readable on a wider array of scanners including those which had only a single, fixed excitation wavelength such as the Affymetrix GCS3000-4C confocal laser scanner. We have explored several solutions to this challenge, including the use of both FRET dye *(18)* and QD conjugates *(19,20)*. Both techniques allow for the creation of a series of dyes where each member is excitable at a common wavelength but emits at successively longer wavelengths and can encode a distinct channel of information (e.g., markers with "A,"- "C,"- "G,"- or "T"-containing SNPs). QDs, whose use we describe here, have the additional advantages of being very photostable and having a broadband absorption spectrum with shorter, more energetic wavelengths producing higher levels of fluorescence emission for a given QD *(21)*. Aside from a few preliminary low-multiplex efforts applying QDs to the detection of nucleic acids on microarrays *(22,23)*, this is the first demonstration of their utility in robust, high-multiplex genomic assays and in multicolor formats.

4.2. Development of QD Conjugates

We have developed a set of four custom QD conjugates (in collaboration with Quantum Dot Corporation, Hayward, CA), emitting within the range of 565 to 705 nm, for staining of the MIP-assay target that has been hybridized to Affymetrix GeneChip Universal Tag arrays. Each QD conjugate reacts with its unique binding partner on the array and thus stains only bound duplex targets from the corresponding channel (i.e., the "A," "C," "G," or "T" reactions). Nonspecific binding is normally quite low and supports the high S/N values characteristic of the MIP assay (e.g., S/N > 100 for all channels in a 12,000-plex reaction; data not shown). Cross-reactivity between the four conjugates is negligible and permits similarly high S/B values, which also factor in biochemical specificity of the assay on top of the S/N calculations. **Figure 2** shows comparative chip views of 5000-plex MIP-genotyping products stained with either the QD or organic dye conjugates.

4.3. Comparison to Organic Dyes

To assess the relative performance of the previous set of organic dye conjugates with their QD counterparts (GeneChip Targeted Genotyping release 1.0), we genotyped 80 Caucasian DNA samples from the HapMap CEPH plate on two separate occasions with the same MIP-genotyping protocol and the same 10,000-plex MIP probe panel (cSNP I, rev2). Both sets of samples were hybridized to GeneChip Universal 10K Tag arrays and one sample set stained with organic dyes, whereas the other stained with QDs. The organic dye-labeled arrays were imaged on a modified ImageExpress 5000A charge-coupled device imager with broadband excitation from a xenon arc lamp and appropriate excitation and emission filters (Molecular Devices, Inc., Union City, CA), whereas those stained with QDs were scanned on the GCS3000-4C confocal laser scanner with a single 532-nm excitation source, and fitted with appropriate emission filters. In early experiments using these same four QD emission wavelengths, individual arrays were scanned on both systems and found to give similar S/N values for each channel (data not shown), suggesting that performance differences in the current experiments are likely attributable to the dyes themselves. The 80 samples included 25 trios (father–mother–child) for determination of Mendelian error (trio concordance) and 5 repeat samples for measurement of repeatability.

The quality control (QC) summary results for these experiments (both analyzed with Affymetrix Software, GTGS v1.5) are shown in **Fig. 3**. These data represent an initial per-chip view (one chip per sample) of several high-level metrics—call rate % (**Fig. 3A**), S/N and S/B (**Fig. 3B**)—prior to clustering analysis. This QC-threshold analysis calls genotypes for each sample individually, and only for those markers that pass minimum S/N and S/B ratios and that fall within a prescribed range of signal ratios for the two signal channels (S_1 and S_2) of a given bi-allelic SNP (e.g., for homozygotes, $S_1/S_2 > 12$ or $S_1/S_2 < 1/12$, and for heterozygotes, $1/3.4 < S_1/S_2 < 3.4$). Samples are further passed or failed according to a number of criteria, including the requirement for successful scoring of at least 80% of the markers in a sample; only passed samples are subsequently used in cluster analysis, which further refines the genotype calls according to clustering of each marker across all samples passed in the initial QC evaluation (*see* **Fig. 4** and **Table 1**). The QC call rates were several percent higher for the QD-stained samples (95.0 ± 1.2%) than for their organic dye-stained counterparts (92.4 ± 1.2%). Similarly, S/N was 17% higher for QDs (129 ± 29 vs 110 ± 13) and S/B was 28% higher for QDs (68 ± 15 vs 53 ± 7). All of these differences in metrics which suggest better performance with QD staining were highly significant by *t*-test ($p < 10^{-6}$). It should be appreciated, however, that all 80 samples in each set passed all QC metrics and gave high-quality genotypes. Note, also, that the QC-threshold analysis was performed using the larger, uncut set of 11,472 markers in the panel than what was used for clustering analysis (*see* **Subheading 4.4.**).

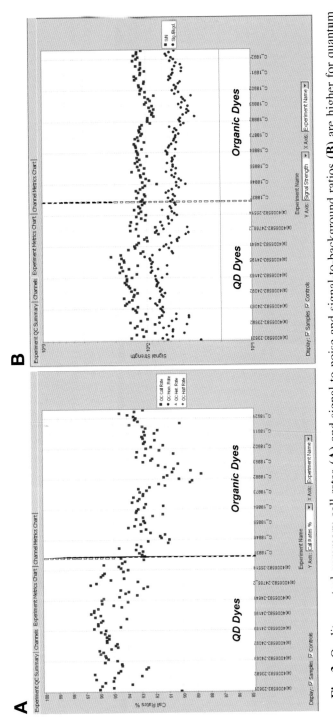

Fig. 3. Quality control summary call rates (**A**) and signal-to-noise and signal-to-background ratios (**B**) are higher for quantum dot-stained than organic dye-stained arrays. This figure shows a production run with each dot representing more than 10,000 genotypes averaged across one sample. The data shown includes about 2 million genotypes.

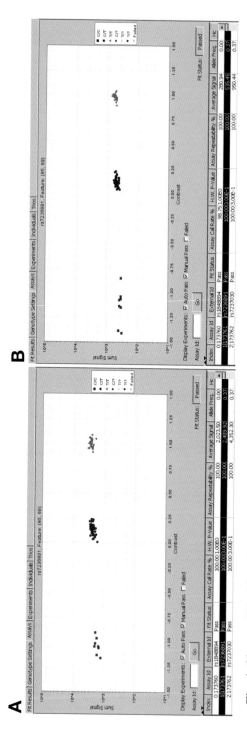

Fig. 4. Clustered HapMap CEPH samples show comparable clustering for single markers across many samples with quantum dots (**A**) and organic dye (**B**) conjugates (10K cSNPI, rev2 assay panel). Red and green clusters represent homozygotes for the major (green) and minor (red) alleles, whereas blue clusters represent individuals heterozygous for this marker.

Table 1
Clustered HapMap CEPH Samples Show Comparable or Better Genotyping Performance With Quantum Dot-Stained Than Organic Dye-Stained Arrays Using 10K cSNPI, rev2 Assay Panel

	QDs	Organic Dyes
# Experiments	80	**80**
# Repeats	5	**5**
# Unique Trios	25	**25**
Total # Assays	9,651	**9,651**
# Passed Assays	9,638	**9,545**
Passed Assays (%)	99.87	**98.90**
Cluster Fit Repeatability (%)	99.92	**99.84**
Trio Concordance (%)	99.79	**99.81**
Completeness (%)	99.25	**99.17**
Passed Assays % x Completeness %	**99.11**	98.08

4.4. Clustering Analysis

A more critical comparison can be seen in the performance metrics of **Table 1**, which derive from clustering analysis of a reduced subset of the 9651 most robust markers in each of these two sample sets. (Similar results were obtained, however, when the analysis was done with the larger set of 11,472 markers; data not shown.) "Passed assays" in cluster analysis refer to only those markers that are called in more than 80% of the samples that passed the QC-summary criteria. In the cluster results shown for 9651 markers, only 13 markers (0.13%) failed to call in at least 80% of the QC'd samples (9638/9651 or 99.9% passed assays), whereas 106 markers (1.1%) failed to call in the organic dye-labeled samples (9545/9651 or 98.9% passed assays). Despite calling approx 1% more markers in the QD-stained samples, the "completeness"—measure of the number of missing calls for all markers scored in a sample set—was nearly identical for the two dye sets (99.25% for QDs vs 99.17% for the organic dyes). Thus, the QD-stained samples delivered approx 1% more genotypes overall (as measured by "passed assays × completeness %" of 99.11% vs 98.08%). Repeatability of the five replicates in each set of 80 samples was slightly higher with the QDs (99.92% vs 99.84%), whereas accuracy, assessed by trio concordance among the 25 mother–father–child trios, was equivalent (99.79% for QDs and 99.81% for the organic dyes). A representative view of homozygous (red and green) and heterozygous (blue) clusters for a single marker is seen in **Fig. 4** for both sample sets, and shows the overall similarity in quality of clusters.

These performance comparisons for the two dye sets are consistent with data from four repetitions with the organic dyes and two repetitions with the QD dyes involving multiple assay panel lots. Thus, the QD conjugates perform at least as well as, and probably somewhat better than, their organic dye counterparts in MIP genotyping insofar as they return more data of comparably high quality. In preliminary experiments, the MIP-genotyping assay has been found to deliver similar performance at the greater than 50,000-plex level using the same QD conjugates.

4.5. Application Beyond Genotyping

Beyond MIP genotyping, however, QD conjugates are also capable of supporting quantitative assays in a multicolor format on microarrays. We have developed a modified, three-color version of the MIP-genotyping assay that enables simultaneous marker copy number and allele ratio determinations *(6)*. To achieve this, a panel containing only C/T markers is scored in two channels ("C" and "T" MIP reactions) against a constant, pooled reference DNA in a third channel. From this data and a set of reference samples used for copy number calibration, the copy number of each locus and its allele ratio can be determined. In **Fig. 5**, we show quantitative performance of the MIP targeted-genotyping assay using diploid cell lines with one, two or four copies of the X chromosome. The data shown is unsmoothed and contains over 10,000 copy number measurements across the human genome. The results indicate that accurate copy number measurements on a genome-wide scale are enabled by the MIP assay.

5. Conclusion

We have shown that QDs can be used beneficially in a commercial, highly multiplexed genotyping assay. The assay produces genotyping results that are superior to organic dyes, and allows the use of simpler instrumentation. In addition, we have demonstrated other applications for QDs in microarray labeling.

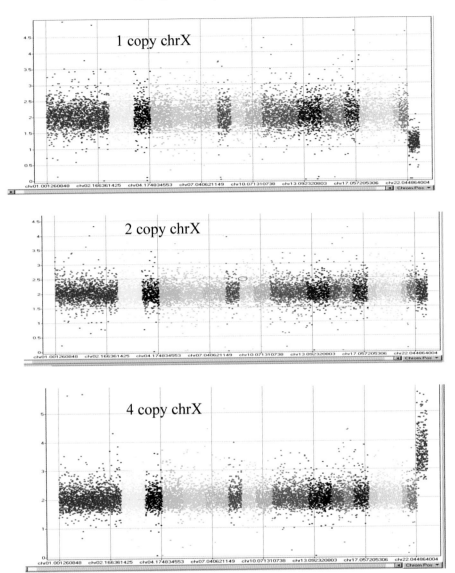

Fig. 5. Quantitative molecular inversion probe genotyping assay with three-color quantum dot staining can accurately determine copy number of genomic DNA with approx two-marker resolution (10K assay panel). Shown are diploid human cell lines with one, two, or four copies of the X-chromosome. Each point represents the copy number (*y*-axis) for a single marker positioned along the chromosomes (*x*-axis). Each chromosome is colored distinctly from its neighboring chromosome, beginning with Chr 1 at the left (in red) through Chr 22 at the right (in peach) and ending with the X-chromosome (in blue) on the extreme right. Even in the unsmoothed data, the difference in copy number is readily apparent.

References

1. Hardenbol, P., Banér, J., Jain, M., et al. (2003) Multiplexed genotyping with sequence-tagged molecular inversion probes. *Nat. Biotechnol.* **21,** 673–678.
2. Matsuzaki, H., Dong, S., Loi, H., et al. (2004) Genotyping over 100,000 SNPs on a pair of oligonucleotide arrays. *Nat. Methods* **1,** 109–111.
3. Shen, R., Fan, J. B., Campbell, D., et al. (2005) High-throughput SNP genotyping on universal bead arrays. *Mutat. Res.* **573,** 70–82.
4. Kidgell, C. and Winzeler, E. A. (2005) Elucidating genetic diversity with oligonucleotide arrays. *Chromosome Res.* **13,** 225–235.
5. Davies, J. J., Wilson, I. M. and Lam, W. L. (2005) Array CGH technologies and their applications to cancer genomes. *Chromosome Res.* **13,** 237–248.
6. Wang, Y., Moorhead, M., Karlin-Neumann, G., et al. (2005) Allele quantification using molecular inversion probes (MIP). *Nucleic Acids Res.* **33,** e183.
7. Schena, M., Heller, R. A., Theriault, T. P., Konrad, K., Lachenmeier, E., and Davis, R. W. (1998) Microarrays: biotechnology's discovery platform for functional genomics. *Trends Biotechnol.* **16,** 301–306.
8. Mockler, T. C., Chan, S., Sundaresan, A., Chen, H., Jacobsen, S.E., and Ecker, J. R. (2005) Applications of DNA tiling arrays for whole-genome analysis. *Genomics* **85,** 1–15.
9. Ren, B. and Dynlacht, B. D. (2004) Use of chromatin immunoprecipitation assays in genome-wide location analysis of mammalian transcription factors. *Methods Enzymol.* **376,** 304–315.
10. Bertone, P., Gerstein, M., and Snyder, M. (2005) Applications of DNA tiling arrays to experimental genome annotation and regulatory pathway discovery. *Chromosome Res.* **13,** 259–274.
11. MacAlpine, D. M. and Bell, S. P. (2005) A genomic view of eukaryotic DNA replication. *Chromosome Res.* **13,** 309–326.
12. Keene, J. D. and Lager, P. J. (2005) Post-transcriptional operons and regulons coordinating gene expression. *Chromosome Res.* **13,** 327–337.
13. van Steensel, B. (2005) Mapping of genetic and epigenetic regulatory networks using microarrays. *Nat. Genet.* **37,** S18–S24.
14. Martienssen, R. A., Doerge, R. W., and Colot, V. (2005) Epigenomic mapping in Arabidopsis using tiling microarrays. *Chromosome Res.* **13,** 299–308.
15. Fredriksson, S., Gullberg, M., Jarvius, J., et al. (2002) Protein detection using proximity-dependent DNA ligation assays. *Nat. Biotechnol.* **20,** 473–477.
16. Gullberg, M., Gústafsdóttir, S. M., Schallmeiner, E., et al. (2004) Cytokine detection by antibody-based proximity ligation. *Proc. Natl. Acad. Sci. USA* **101,** 8420–8424.
17. Hardenbol, P., Yu, F., Belmont, J., et al. (2005) Highly multiplexed molecular inversion probe genotyping: Over 10,000 targeted SNPs genotyped in a single tube assay. *Genome Res.* **15,** 269–275.
18. Berlier, J. E., Rothe, A., Buller, G., et al. (2003) Quantitative comparison of long-wavelength Alexa Fluor dyes to Cy dyes: fluorescence of the dyes and their bioconjugates. *J. Histochem. Cytochem.* **51,** 1699–1712.

19. Chan, W. C., Maxwell, D. J., Gao, X., Bailey, R. E., Han, M., and Nie, S. (2002) Luminescent quantum dots for multiplexed biological detection and imaging. *Curr. Opin. Biotechnol.* **13**, 40–46.
20. Wu, X., Liu, H., Liu, J., et al. (2003) Immunofluorescent labeling of cancer marker Her2 and other cellular targets with semiconductor quantum dots. *Nat. Biotechnol.* **21**, 41–46.
21. Michalet, X., Pinaud, F. F., Bentolila, L. A., et al. (2005) Quantum dots for live cells, in vivo imaging, and diagnostics. *Science* **307**, 538–544.
22. Gerion, D., Chen, F., Kannan, B., et al. (2003) Room-temperature single-nucleotide polymorphism and multiallele DNA detection using fluorescent nanocrystals and microarrays. *Anal. Chem.* **75**, 4766–4772.
23. Liang, R.-Q., Li, W., Li, Y., et al. (2005) An oligonucleotide microarray for microRNA expression analysis based on labeling RNA with quantum dot and nanogold probe. *Nucleic Acids Res.* **33**, e17.

Index